STUDENT SOLUTIONS MANUAL
to Accompany

SALAS AND HILLE'S
CALCULUS
SEVERAL VARIABLES
Chapters 12-17

Seventh Edition

REVISED BY
GARRET J. ETGEN

John Wiley & Sons, Inc.
New York • Chichester • Brisbane • Toronto • Singapore

ISBN 0-471-17212-X

Printed in the United States of America

10 9 8 7 6 5 4 3 2 1

Printed and bound by Malloy Lithographing, Inc.

CONTENTS

CHAPTER 12

SECTION 12.1

1.

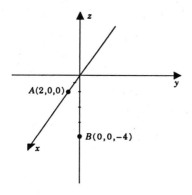

length $\overline{AB}$: $2\sqrt{5}$

midpoint: $(1, 0, -2)$

3.

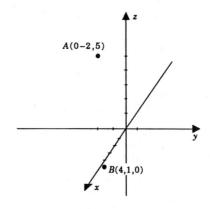

length $\overline{AB}$: $5\sqrt{2}$

midpoint: $\left(2, -\frac{1}{2}, \frac{5}{2}\right)$

5.

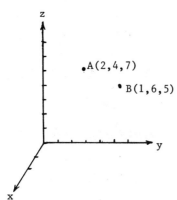

length $\overline{AB}$: 3 midpoint: $\left(\frac{3}{2}, 5, 6\right)$

7. $z = -2$ **9.** $y = 1$ **11.** $x = 3$

13. $x^2 + (y-2)^2 + (z+1)^2 = 9$ **15.** $(x-2)^2 + (y-4)^2 + (z+4)^2 = 36$

17. $(x-3)^2 + (y-2)^2 + (z-2)^2 = 13$ **19.** $(x-2)^2 + (y-3)^2 + (z+4)^2 = 25$

21.

$$x^2 + y^2 + z^2 + 4x - 8y - 2z + 5 = 0$$

$$x^2 + 4x + 4 + y^2 - 8y + 16 + z^2 - 2z + 1 = -5 + 4 + 16 + 1$$

$$(x+2)^2 + (y-4)^2 + (z-1)^2 = 16$$

center: $(-2, 4, 1)$,   radius: 4

23.
$$2x^2 + 2y^2 + 2z^2 + 8x - 4y = -1$$
$$2(x^2 + 4x + 4) + 2(y^2 - 2y + 1) + 2z^2 = -1 + 8 + 2$$
$$(x + 2)^2 + (y - 1)^2 + z^2 = \frac{9}{2}$$

center: $(-2, 1, 0)$,   radius: $\frac{3}{2}\sqrt{2}$

25. $(2, 3, -5)$ **27.** $(-2, 3, 5)$ **29.** $(-2, 3, -5)$

31. $(-2, -3, -5)$ **33.** $(2, -5, 5)$ **35.** $(-2, 1, -3)$

37. Each such sphere has an equation of the form
$$(x - a)^2 + (y - a)^2 + (z - a)^2 = a^2.$$

Substituting $x = 5$, $y = 1$, $z = 4$ we get
$$(5 - a)^2 + (1 - a)^2 + (4 - a)^2 = a^2.$$

This reduces to $a^2 - 10a + 21 = 0$ and gives $a = 3$ or $a = 7$. The equations are:
$$(x - 3)^2 + (y - 3)^2 + (z - 3)^2 = 9; \quad (x - 7)^2 + (y - 7)^2 + (z - 7)^2 = 49$$

39. Not a sphere; this equation is equivalent to:
$$(x - 2)^2 + (y + 2)^2 + (z + 3)^2 = -3$$

which has no (real) solutions.

41. $d(PR) = \sqrt{14}$, $d(QR) = \sqrt{45}$, $d(PQ) = \sqrt{59}$; $[d(PR)]^2 + [d(QR)]^2 = [d(PQ)]^2$

43. (a) Take R as (x, y, z). Since
$$d(P, R) = t\, d(P, Q)$$
we conclude by similar triangles that
$$d(AR) = t\, d(B, Q)$$
and therefore
$$z - a_3 = t(b_3 - a_3).$$
Thus
$$z = a_3 + t(b_3 - a_3).$$

In similar fashion

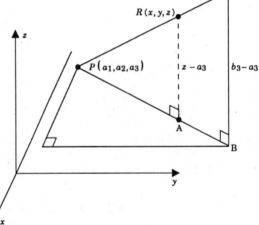

$$x = a_1 + t(b_1 - a_1) \quad \text{and} \quad y = a_2 + t(b_2 - a_2).$$

(b) The midpoint of PQ, $\left(\dfrac{a_1 + b_1}{2}, \dfrac{a_2 + b_2}{2}, \dfrac{a_3 + b_3}{2} \right)$,   occurs at $t = \dfrac{1}{2}$.

SECTION 12.2

1. $\overrightarrow{PQ} = (3, 4, -2);$ $\|\overrightarrow{PQ}\| = \sqrt{29}$

3. $\overrightarrow{PQ} = (-2, 6);$ $\|\overrightarrow{PQ}\| = 2\sqrt{10}$

5. $\overrightarrow{PQ}(-4, 2, 2);$ $\|\overrightarrow{PQ}\| = 2\sqrt{6}$

7. $2\mathbf{a} - \mathbf{b} = (2 \cdot 1 - 3, 2 \cdot [-2] - 0, 2 \cdot 3 + 1)$
$$= (-1, -4, 7)$$

9. $-2\mathbf{a} + \mathbf{b} - \mathbf{c} = [-2(\mathbf{a} - \mathbf{b})] - \mathbf{c} = (1 + 4, 4 - 2, -7 - 1) = (5, 2, -8)$

11. $3\mathbf{i} - 4\mathbf{j} + 6\mathbf{k}$

13. $-3\mathbf{i} - \mathbf{j} + 8\mathbf{k}$

15. 5

17. 3

19. $\sqrt{6}$

21. (a) $\mathbf{a}$, $\mathbf{c}$, and $\mathbf{d}$ since $\mathbf{a} = \frac{1}{3}\mathbf{c} = -\frac{1}{2}\mathbf{d}$

    (b) $\mathbf{a}$ and $\mathbf{c}$ since $\mathbf{a} = \frac{1}{3}\mathbf{c}$

    (c) $\mathbf{a}$ and $\mathbf{c}$ both have direction opposite to $\mathbf{d}$

23. $\|\mathbf{a}\| = 5;$ $\dfrac{\mathbf{a}}{\|\mathbf{a}\|} = \left(\dfrac{3}{5}, -\dfrac{4}{5}\right)$

25. $\|\mathbf{a}\| = 5;$ $\dfrac{\mathbf{a}}{\|\mathbf{a}\|} = \left(-\dfrac{4}{5}, 0, \dfrac{3}{5}\right)$

27. $\|\mathbf{a}\| = 3;$ $\dfrac{\mathbf{a}}{\|\mathbf{a}\|} = \dfrac{1}{3}\mathbf{i} - \dfrac{2}{3}\mathbf{j} + \dfrac{2}{3}\mathbf{k}$

29. $\|\mathbf{a}\| = \sqrt{14};$ $-\dfrac{\mathbf{a}}{\|\mathbf{a}\|} = \dfrac{1}{\sqrt{14}}\mathbf{i} - \dfrac{3}{\sqrt{14}}\mathbf{j} - \dfrac{2}{\sqrt{14}}\mathbf{k}$

31. (i) $\mathbf{a} + \mathbf{b}$ (ii) $-(\mathbf{a} + \mathbf{b})$ (iii) $\mathbf{a} - \mathbf{b}$ (iv) $\mathbf{b} - \mathbf{a}$

33. (a) $\mathbf{a} - 3\mathbf{b} + 2\mathbf{c} + 4\mathbf{d} = (2\mathbf{i} - \mathbf{k}) - 3(\mathbf{i} + 3\mathbf{j} + 5\mathbf{k}) + 2(-\mathbf{i} + \mathbf{j} + \mathbf{k}) + 4(\mathbf{i} + \mathbf{j} + 6\mathbf{k})$
$$= \mathbf{i} - 3\mathbf{j} + 10\mathbf{k}$$

    (b) The vector equation
$$(1, 1, 6) = A(2, 0, -1) + B(1, 3, 5) + C(-1, 1, 1)$$

    implies
$$\begin{array}{rcrcrcr} 1 & = & 2A & + & B & - & C, \\ 1 & = & & & 3B & + & C, \\ 6 & = & -A & + & 5B & + & C. \end{array}$$

    Simultaneous solution gives $A = -2,$ $B = \frac{3}{2},$ $C = -\frac{7}{2}.$

35. $\|3\mathbf{i} + \mathbf{j}\| = \|\alpha\mathbf{j} - \mathbf{k}\|$ $\Longrightarrow$ $10 = \alpha^2 + 1$ so $\alpha = \pm 3$

37. $\|\alpha\mathbf{i} + (\alpha - 1)\mathbf{j} + (\alpha + 1)\mathbf{k}\| = 2$ $\Longrightarrow$ $\alpha^2 + (\alpha - 1)^2 + (\alpha + 1)^2 = 4$
$$\Longrightarrow \quad 3\alpha^2 = 2 \quad \text{so} \quad \alpha = \pm\frac{1}{3}\sqrt{6}$$

39. $\pm\frac{2}{13}\sqrt{13}\,(3\mathbf{j} + 2\mathbf{k})$ since $\|\alpha(3\mathbf{j} + 2\mathbf{k})\| = 2$ $\Longrightarrow$ $\alpha = \pm\frac{2}{13}\sqrt{13}$

41. $\mathbf{v} = (2\cos 30°)\,\mathbf{i} + (2\sin 30°)\,\mathbf{j} = \sqrt{3}\,\mathbf{i} + \mathbf{j}$

43. $\mathbf{v} = \cos(\pi/4)\,\mathbf{i} + \sin(\pi/4)\,\mathbf{j} = \dfrac{\sqrt{2}}{2}\,\mathbf{i} + \dfrac{\sqrt{2}}{2}\,\mathbf{j}$

45. Since the $\mathbf{i}$ component is twice the $\mathbf{j}$ component, $\mathbf{v} = 2y\,\mathbf{i} + y\,\mathbf{j}$. Now, $\|\mathbf{v}\| = \sqrt{4y^2 + y^2} = 3$ which implies that $y = \dfrac{3}{\sqrt{5}}$. Thus, $\mathbf{v} = \dfrac{6}{\sqrt{5}}\,\mathbf{i} + \dfrac{3}{\sqrt{5}}\,\mathbf{j}$ or $\mathbf{v} = -\dfrac{6}{\sqrt{5}}\,\mathbf{i} - \dfrac{3}{\sqrt{5}}\,\mathbf{j}$.

47. If $\mathbf{a}$ and $\mathbf{b}$ are the sides of a triangle, then $\mathbf{b} - \mathbf{a}$ is the third side. Now $\|\mathbf{a}\| = \sqrt{2^2 + (-1)^2} = \sqrt{5}$, $\|\mathbf{b}\| = \sqrt{1^2 + 2^2} = \sqrt{5}$, and $\|\mathbf{b} - \mathbf{a}\| = \sqrt{(1-2)^2 + (2+1)^2} = \sqrt{10}$. The triangle is a right triangle since $\|\mathbf{a}\|^2 + \|\mathbf{b}\|^2 = \|\mathbf{b} - \mathbf{a}\|^2$.

49. (a) Since $\|\mathbf{a} - \mathbf{b}\|$ and $\|\mathbf{a} + \mathbf{b}\|$ are the lengths of the diagonals of the parallelogram, the parallelogram must be a rectangle.

  (b) Simplify
  $$\sqrt{(a_1 - b_1)^2 + (a_2 - b_2)^2 + (a_3 - b_3)^2} = \sqrt{(a_1 + b_1)^2 + (a_2 + b_2)^2 + (a_3 + b_3)^2}.$$

51. (a)
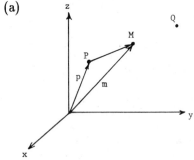

  (b) Let $P = (x_1, y_1, z_1)$, $Q = (x_2, y_2, z_2)$, and
  $M = (x_m, y_m, z_m)$. Then

  $(x_m, y_m, z_m) = (x_1, y_1, z_1) + \tfrac{1}{2}(x_2 - x_1, y_2 - y_1, z_2 - z_1)$

  $= \left(\dfrac{x_1 + x_2}{2}, \dfrac{y_1 + y_2}{2}, \dfrac{z_1 + z_2}{2} \right)$

53. $\|\mathbf{F}_1\|\sin 40° + \|\mathbf{F}_2\|\sin 25° = 200$ and $\|\mathbf{F}_1\|\cos 40° = \|\mathbf{F}_2\|\cos 25°$

  $\Longrightarrow$ $\|\mathbf{F}_1\| = 200.02$ and $\|\mathbf{F}_2\| = 169.05$

  $\mathbf{F}_1 = -\|\mathbf{F}_1\|\cos 40°\,\mathbf{i} + \|\mathbf{F}_1\|\sin 40°\,\mathbf{j} = -153.21\,\mathbf{i} + 128.56\,\mathbf{j}$
  $\mathbf{F}_2 = \|\mathbf{F}_2\|\cos 25°\,\mathbf{i} + \|\mathbf{F}_2\|\sin 25°\,\mathbf{j} = 153.21\,\mathbf{i} + 71.44\,\mathbf{j}$

55. $\mathbf{V}_1 = 600\sin 30°\,\mathbf{i} + 600\cos 30°\,\mathbf{j} = 300\,\mathbf{i} + 300\sqrt{3}\,\mathbf{j}$ and
  $\mathbf{V}_2 = 50\sin 45°\,\mathbf{i} - 50\cos 45°\,\mathbf{j} = 25\sqrt{2}\,\mathbf{i} - 25\sqrt{2}\,\mathbf{j}$

  $\mathbf{V} = \mathbf{V}_1 + \mathbf{V}_2 = (300 + 25\sqrt{2})\,\mathbf{i} + (300\sqrt{3} - 25\sqrt{2})\,\mathbf{j} \cong 335.36\,\mathbf{i} + 484.26\,\mathbf{j}$

  true course: $\theta = \tan^{-1}\dfrac{335.36}{484.26} = 34.70°$;   or $N\,34.70°\,E$.

  ground speed: $\|\mathbf{V}\| = \sqrt{(335.56)^2 + (484.26)^2} \cong 589.05$ mi/hr

57. (a) $\|\mathbf{r} - \mathbf{a}\| = 3$ where $\mathbf{a} = a_1\mathbf{i} + a_2\mathbf{j} + a_3\mathbf{k}$
  (b) $\|\mathbf{r}\| \le 2$ (c) $\|\mathbf{r} - \mathbf{a}\| \le 1$ where $\mathbf{a} = a_1\mathbf{i} + a_2\mathbf{j} + a_3\mathbf{k}$
  (d) $\|\mathbf{r} - \mathbf{a}\| = \|\mathbf{r} - \mathbf{b}\|$ (e) $\|\mathbf{r} - \mathbf{a}\| + \|\mathbf{r} - \mathbf{b}\| = k$

SECTION 12.3

1. $\mathbf{a} \cdot \mathbf{b} = (2)(-2) + (-3)(0) + (1)(3) = -1$ 3. $\mathbf{a} \cdot \mathbf{b} = (2)(1) + (-4)(1/2) = 0$

5. $\mathbf{a} \cdot \mathbf{b} = (2)(1) + (1)(1) - (2)(2) = -1$ 7. $\mathbf{a} \cdot \mathbf{b}$

9. $(\mathbf{a} - \mathbf{b}) \cdot \mathbf{c} + \mathbf{b} \cdot (\mathbf{c} + \mathbf{a}) = \mathbf{a} \cdot \mathbf{c} - \mathbf{b} \cdot \mathbf{c} + \mathbf{b} \cdot \mathbf{c} + \mathbf{b} \cdot \mathbf{a} = \mathbf{a} \cdot (\mathbf{b} + \mathbf{c})$

11. (a) $\mathbf{a} \cdot \mathbf{b} = (2)(3) + (1)(-1) + (0)(2) = 5$

     $\mathbf{a} \cdot \mathbf{c} = (2)(4) + (1)(0) + (0)(3) = 8$

     $\mathbf{b} \cdot \mathbf{c} = (3)(4) + (-1)(0) + (2)(3) = 18$

     (b) $\|\mathbf{a}\| = \sqrt{5}$, $\|\mathbf{b}\| = \sqrt{14}$, $\|\mathbf{c}\| = 5$. Then,
     $$\cos \sphericalangle(\mathbf{a},\mathbf{b}) = \frac{\mathbf{a} \cdot \mathbf{b}}{\|\mathbf{a}\| \, \|\mathbf{b}\|} = \frac{5}{(\sqrt{5})\,(\sqrt{14})} = \frac{1}{14}\sqrt{70},$$
     $$\cos \sphericalangle(\mathbf{a},\mathbf{c}) = \frac{8}{(\sqrt{5})\,(5)} = \frac{8}{25}\sqrt{5},$$
     $$\cos \sphericalangle(\mathbf{b},\mathbf{c}) = \frac{18}{(\sqrt{14})\,(5)} = \frac{9}{35}\sqrt{14}.$$

     (c) $\mathbf{u_b} = \frac{1}{\sqrt{14}}(3\mathbf{i} - \mathbf{j} + 2\mathbf{k})$, $\mathrm{comp_b}\,\mathbf{a} = \mathbf{a} \cdot \mathbf{u_b} = \frac{1}{\sqrt{14}}(6-1) = \frac{5}{14}\sqrt{14}$,

     $\mathbf{u_c} = \frac{1}{5}(4\mathbf{i} + 3\mathbf{k})$, $\mathrm{comp_c}\,\mathbf{a} = \mathbf{a} \cdot \mathbf{u_c} = \frac{8}{5}$

     (d) $\mathbf{proj_b}\,\mathbf{a} = (\mathrm{comp_b}\,\mathbf{a})\,\mathbf{u_b} = \frac{5}{14}(3\mathbf{i} - \mathbf{j} + 2\mathbf{k})$

     $\mathbf{proj_c}\,\mathbf{a} = (\mathrm{comp_c}\,\mathbf{a})\,\mathbf{u_c} = \frac{8}{25}(4\mathbf{i} + 3\mathbf{k})$

13. $\mathbf{u} = \cos\frac{\pi}{3}\mathbf{i} + \cos\frac{\pi}{4}\mathbf{j} + \cos\frac{2\pi}{3}\mathbf{k} = \frac{1}{2}\mathbf{i} + \frac{1}{2}\sqrt{2}\mathbf{j} - \frac{1}{2}\mathbf{k}$

15. $\cos\theta = \frac{(3\mathbf{i} - \mathbf{j} - 2\mathbf{k}) \cdot (\mathbf{i} + 2\mathbf{j} - 3\mathbf{k})}{\|3\mathbf{i} - \mathbf{j} - 2\mathbf{k}\|\,\|\mathbf{i} + 2\mathbf{j} - 3\mathbf{k}\|} = \frac{7}{\sqrt{14}\sqrt{14}} = \frac{1}{2}$, $\theta = \frac{\pi}{3}$

17. Since $\|\mathbf{i} - \mathbf{j} + \sqrt{2}\,\mathbf{k}\| = 2$, we have $\cos\alpha = \frac{1}{2}$, $\cos\beta = -\frac{1}{2}$, $\cos\gamma = \frac{1}{2}\sqrt{2}$.
    The direction angles are $\frac{1}{3}\pi$, $\frac{2}{3}\pi$, $\frac{1}{4}\pi$.

19. $\theta = \cos^{-1}\frac{\mathbf{a} \cdot \mathbf{b}}{\|\mathbf{a}\|\|\mathbf{b}\|} = \cos^{-1}\left(\frac{-1}{\sqrt{231}}\right) \cong 2.2$ radians or $126.3°$

21. $\theta = \cos^{-1}\frac{\mathbf{a} \cdot \mathbf{b}}{\|\mathbf{a}\|\|\mathbf{b}\|} = \cos^{-1}\left(\frac{-13}{5\sqrt{10}}\right) \cong 2.5$ radians or $145.3°$

23. $\|\mathbf{a}\| = \sqrt{1^2 + 2^2 + 2^2} = 3$; $\cos\alpha = \frac{1}{3}$, $\cos\beta = \frac{2}{3}$, $\cos\gamma = \frac{2}{3}$

     $\alpha \cong 70.5°$ $\beta \cong 48.2°$, $\gamma \cong 48.2°$

25. $\|\mathbf{a}\| = \sqrt{3^2 + (12)^2 + 4^2} = 13$; $\cos\alpha = \frac{3}{13}$, $\cos\beta = \frac{12}{13}$ $\cos\gamma = \frac{4}{13}$

     $\alpha \cong 76.7°$ $\beta \cong 22.6°$, $\gamma \cong 72.1°$

27.   (a) $\mathbf{proj_b}\,\alpha\mathbf{a} = (\alpha\mathbf{a} \cdot \mathbf{u_b})\mathbf{u_b} = \alpha(\mathbf{a} \cdot \mathbf{u_b})\mathbf{u_b} = \alpha\,\mathbf{proj_b}\,\mathbf{a}$

   (b) $$\mathbf{proj_b}\,(\mathbf{a} + \mathbf{c}) = [(\mathbf{a} + \mathbf{c}) \cdot \mathbf{u_b}]\,\mathbf{u_b}$$

   $$= (\mathbf{a} \cdot \mathbf{u_b} + \mathbf{c} \cdot \mathbf{u_b})\mathbf{u_b}$$

   $$= (\mathbf{a} \cdot \mathbf{u_b})\mathbf{u_b} + (\mathbf{c} \cdot \mathbf{u_b})\mathbf{u_b} = \mathbf{proj_b}\,\mathbf{a} + \mathbf{proj_b}\,\mathbf{c}$$

29.   (a)   For $\mathbf{a} \neq \mathbf{0}$ the following statements are equivalent:

   $$\mathbf{a} \cdot \mathbf{b} = \mathbf{a} \cdot \mathbf{c}, \quad \mathbf{b} \cdot \mathbf{a} = \mathbf{c} \cdot \mathbf{a},$$

   $$\mathbf{b} \cdot \frac{\mathbf{a}}{\|\mathbf{a}\|} = \mathbf{c} \cdot \frac{\mathbf{a}}{\|\mathbf{a}\|}, \quad \mathbf{b} \cdot \mathbf{u_a} = \mathbf{c} \cdot \mathbf{u_a}$$

   $$(\mathbf{b} \cdot \mathbf{u_a})\mathbf{u_a} = (\mathbf{c} \cdot \mathbf{u_a})\mathbf{u_a},$$

   $$\mathbf{proj_a}\,\mathbf{b} = \mathbf{proj_a}\,\mathbf{c}.$$

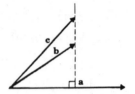

   $$\mathbf{a} \cdot \mathbf{b} = \mathbf{a} \cdot \mathbf{c} \quad \text{but} \quad \mathbf{b} \neq \mathbf{c}$$

   (b) $\mathbf{b} = (\mathbf{b} \cdot \mathbf{i})\mathbf{i} + (\mathbf{b} \cdot \mathbf{j})\mathbf{j} + (\mathbf{b} \cdot \mathbf{k})\mathbf{k} = (\mathbf{c} \cdot \mathbf{i})\mathbf{i} + (\mathbf{c} \cdot \mathbf{j})\mathbf{j} + (\mathbf{c} \cdot \mathbf{k})\mathbf{k} = \mathbf{c}$
   $\qquad \underset{\text{(12.3.13)}}{\underline{\quad\quad}} \qquad\qquad\qquad\qquad \text{(12.3.13)}\underline{\quad\quad}$

31.   (a) $\|\mathbf{a} + \mathbf{b}\|^2 - \|\mathbf{a} - \mathbf{b}\|^2 = (\mathbf{a} + \mathbf{b}) \cdot (\mathbf{a} + \mathbf{b}) - (\mathbf{a} - \mathbf{b}) \cdot (\mathbf{a} - \mathbf{b})$
   $$= [(\mathbf{a} \cdot \mathbf{a}) + 2(\mathbf{a} \cdot \mathbf{b}) + (\mathbf{b} \cdot \mathbf{b})] - [(\mathbf{a} \cdot \mathbf{a}) - 2(\mathbf{a} \cdot \mathbf{b}) + (\mathbf{b} \cdot \mathbf{b})] = 4(\mathbf{a} \cdot \mathbf{b})$$

   (b)   The following statements are equivalent:

   $$\mathbf{a} \perp \mathbf{b}, \quad \mathbf{a} \cdot \mathbf{b} = 0, \quad \|\mathbf{a} + \mathbf{b}\|^2 - \|\mathbf{a} - \mathbf{b}\|^2 = 0, \quad \|\mathbf{a} + \mathbf{b}\| = \|\mathbf{a} - \mathbf{b}\|.$$

   (c)   By (b), the relation $\|\mathbf{a} + \mathbf{b}\| = \|\mathbf{a} - \mathbf{b}\|$ gives $\mathbf{a} \perp \mathbf{b}$. The relation $\mathbf{a} + \mathbf{b} \perp \mathbf{a} - \mathbf{b}$ gives

   $$0 = (\mathbf{a} + \mathbf{b}) \cdot (\mathbf{a} - \mathbf{b}) = \|\mathbf{a}\|^2 - \|\mathbf{b}\|^2 \quad \text{and thus} \quad \|\mathbf{a}\| = \|\mathbf{b}\|.$$

   The parallelogram is a square since it has two adjacent sides of equal length and these meet at right angles.

33. $\|\mathbf{a} + \mathbf{b}\|^2 = (\mathbf{a} + \mathbf{b}) \cdot (\mathbf{a} + \mathbf{b}) = \mathbf{a} \cdot \mathbf{a} + 2\mathbf{a} \cdot \mathbf{b} + \mathbf{b} \cdot \mathbf{b} = \|\mathbf{a}\|^2 + 2\mathbf{a} \cdot \mathbf{b} + \|\mathbf{b}\|^2$
   $\|\mathbf{a} - \mathbf{b}\|^2 = (\mathbf{a} - \mathbf{b}) \cdot (\mathbf{a} - \mathbf{b}) = \mathbf{a} \cdot \mathbf{a} - 2\mathbf{a} \cdot \mathbf{b} + \mathbf{b} \cdot \mathbf{b} = \|\mathbf{a}\|^2 - 2\mathbf{a} \cdot \mathbf{b} + \|\mathbf{b}\|^2$

   Add the two equations and the result follows.

35.   Let $\theta_1, \theta_2, \theta_3$ be the direction angles of $-\mathbf{a}$. Then

   $$\theta_1 = \cos^{-1}\left[\frac{(-\mathbf{a} \cdot \mathbf{i})}{\|-\mathbf{a}\|}\right] = \cos^{-1}\left[-\frac{(\mathbf{a} \cdot \mathbf{i})}{\|\mathbf{a}\|}\right] = \cos^{-1}(-\cos\alpha) = \cos^{-1}(\pi - \alpha) = \pi - \alpha.$$

   Similarly $\theta_2 = \pi - \beta$ and $\theta_3 = \pi - \gamma$.

37.   If $\mathbf{a} \perp \mathbf{b}$ and $\mathbf{a} \perp \mathbf{c}$,   then

   $$\mathbf{a} \cdot \mathbf{b} = 0, \quad \mathbf{a} \cdot \mathbf{c} = 0$$

   $$\mathbf{a} \cdot (\alpha\mathbf{b} + \beta\mathbf{c}) = \alpha(\mathbf{a} \cdot \mathbf{b}) + \beta(\mathbf{a} \cdot \mathbf{c}) = 0$$

   $$\mathbf{a} \perp (\alpha\mathbf{b} + \beta\mathbf{c}).$$

39. Existence of decomposition:

$$\mathbf{a} = (\mathbf{a} \cdot \mathbf{u_b})\mathbf{u_b} + [\mathbf{a} - (\mathbf{a} \cdot \mathbf{u_b})\mathbf{u_b}].$$

Uniqueness of decomposition: suppose that

$$\mathbf{a} = \mathbf{a}_{\parallel} + \mathbf{a}_{\perp} = \mathbf{A}_{\parallel} + \mathbf{A}_{\perp}.$$

Then the vector $\mathbf{a}_{\parallel} - \mathbf{A}_{\parallel} = \mathbf{A}_{\perp} - \mathbf{a}_{\perp}$ is both parallel to $\mathbf{b}$ and perpendicular to $\mathbf{b}$. (Exercises 37 and 38.) Therefore it is zero. Consequently $\mathbf{A}_{\parallel} = \mathbf{a}_{\parallel}$ and $\mathbf{A}_{\perp} = \mathbf{a}_{\perp}$.

41. $\cos \dfrac{\pi}{3} = \dfrac{\mathbf{c} \cdot \mathbf{d}}{\|\mathbf{c}\| \, \|\mathbf{d}\|}, \quad \dfrac{1}{2} = \dfrac{2x+1}{x^2+2}, \quad x^2 = 4x; \quad x = 0, 4$

43.

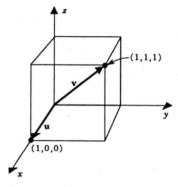

We take $\mathbf{u} = \mathbf{i}$ as an edge and $\mathbf{v} = \mathbf{i} + \mathbf{j} + \mathbf{k}$ as a diagonal of a cube. Then,

$$\cos \theta = \frac{\mathbf{u} \cdot \mathbf{v}}{\|\mathbf{u}\| \, \|\mathbf{v}\|} = \frac{1}{3}\sqrt{3},$$

$$\theta = \cos^{-1}\left(\tfrac{1}{3}\sqrt{3}\right) \cong 0.96 \text{ radians.}$$

45. (a) The direction angles of a vector always satisfy

$$\cos^2 \alpha + \cos^2 \beta + \cos^2 \gamma = 1$$

and, as you can check,

$$\cos^2 \tfrac{1}{4}\pi + \cos^2 \tfrac{1}{6}\pi + \cos^2 \tfrac{2}{3}\pi \neq 1.$$

(b) The relation

$$\cos^2 \alpha + \cos^2 \tfrac{1}{4}\pi + \cos^2 \tfrac{1}{4}\pi = 1$$

gives

$$\cos^2 \alpha + \tfrac{1}{2} + \tfrac{1}{2} = 1, \quad \cos \alpha = 0, \quad a_1 = \|\mathbf{a}\| \cos \alpha = 0.$$

47. Set $\mathbf{u} = a\mathbf{i} + b\mathbf{j} + c\mathbf{k}$. The relations

$$(a\mathbf{i} + b\mathbf{j} + c\mathbf{k}) \cdot (\mathbf{i} + 2\mathbf{j} + \mathbf{k}) = 0 \quad \text{and} \quad (a\mathbf{i} + b\mathbf{j} + c\mathbf{k}) \cdot (3\mathbf{i} - 4\mathbf{j} + 2\mathbf{k}) = 0$$

give

$$a + 2b + c = 0 \qquad 3a - 4b + 2c = 0$$

so that $b = \tfrac{1}{8}a$ and $c = -\tfrac{5}{4}a$.

Then, since $\mathbf{u}$ is a unit vector,

$$a^2 + b^2 + c^2 = 1, \quad a^2 + \left(\frac{a}{8}\right)^2 + \left(\frac{-5a}{4}\right)^2 = 1, \quad \frac{165}{64}a^2 = 1.$$

Thus, $\quad a = \pm\dfrac{8}{165}\sqrt{165} \quad$ and $\quad \mathbf{u} = \pm\dfrac{\sqrt{165}}{165}(8\mathbf{i} + \mathbf{j} - 10\mathbf{k}).$

49.   Place center of sphere at the origin.

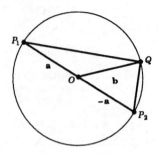

$$\overrightarrow{P_1Q} \cdot \overrightarrow{P_2Q} = (-\mathbf{a} + \mathbf{b}) \cdot (\mathbf{a} + \mathbf{b})$$

$$= -\|\mathbf{a}\|^2 + \|\mathbf{b}\|^2$$

$$= 0.$$

51.   (a) $W = \mathbf{F} \cdot \mathbf{r}$ (b)   0   (c) $\|\mathbf{F}\| \mathbf{i} \cdot (b - a)\mathbf{i} = \|\mathbf{F}\|(b - a)$

53.   (a) $W = (15 \cos 35° \, \mathbf{i} + 15 \sin 35° \, \mathbf{j}) \cdot (50\,\mathbf{i}) = 15 \cdot \cos 35° \cdot 50 = 614.36$ joules

      (b) $W = (15 \cos 50° \, \mathbf{i} + 15 \sin 50° \, \mathbf{j}) \cdot (50 \cos 15° \, \mathbf{i} + 50 \sin 15° \, \mathbf{j})$

$$= 15 \cdot 50(\cos 50° \cos 15° + \sin 50° \sin 15° = 15 \cdot 50 \cos 35° = 614.36 \text{ joules}$$

55.   Let $\|\mathbf{F}_1\| = \|\mathbf{F}_2\| = C.$

      (a) $W_1 = C \cos \theta_1; \quad W_2 = C \cos \theta_2 = C \cos(-\theta_1) = C \cos \theta_1 = W_1 \quad$ Thus, $W_1 = W_2$

      (b) $W_1 = C \cdot \cos(\pi/3) \cdot \|\mathbf{r}\| = \frac{1}{2}C\|\mathbf{r}\|$ and $W_2 = C \cdot \cos(\pi/6) \cdot \|\mathbf{r}\| = \frac{\sqrt{3}}{2}C\|\mathbf{r}\|$

      Thus, $W_2 = \sqrt{3}\, W_1$

SECTION 12.4

1. $(\mathbf{i} + \mathbf{j}) \times (\mathbf{i} - \mathbf{j}) = [\mathbf{i} \times (\mathbf{i} - \mathbf{j})] + [\mathbf{j} \times (\mathbf{i} - \mathbf{j})] = (\mathbf{0} - \mathbf{k}) + (-\mathbf{k} - \mathbf{0}) = -2\mathbf{k}$

3. $(\mathbf{i} - \mathbf{j}) \times (\mathbf{j} - \mathbf{k}) = [\mathbf{i} \times (\mathbf{j} - \mathbf{k})] - [\mathbf{j} \times (\mathbf{j} - \mathbf{k})] = (\mathbf{j} + \mathbf{k}) - (\mathbf{0} - \mathbf{i}) = \mathbf{i} + \mathbf{j} + \mathbf{k}$

5. $(2\mathbf{j} - \mathbf{k}) \times (\mathbf{i} - 3\mathbf{j}) = [2\mathbf{j} \times (\mathbf{i} - 3\mathbf{j})] - [\mathbf{k} \times (\mathbf{i} - 3\mathbf{j})] = (-2\mathbf{k}) - (\mathbf{j} + 3\mathbf{i}) = -3\mathbf{i} - \mathbf{j} - 2\mathbf{k}$

    or

$$(2\mathbf{j} - \mathbf{k}) \times (\mathbf{i} - 3\mathbf{j}) = \begin{vmatrix} \mathbf{i} & \mathbf{j} & \mathbf{k} \\ 0 & 2 & -1 \\ 1 & -3 & 0 \end{vmatrix} = \mathbf{i}\begin{vmatrix} 2 & -1 \\ -3 & 0 \end{vmatrix} - \mathbf{j}\begin{vmatrix} 0 & -1 \\ 1 & -3 \end{vmatrix} + \mathbf{k}\begin{vmatrix} 0 & 2 \\ 1 & -3 \end{vmatrix} = -3\mathbf{i} - \mathbf{j} - 2\mathbf{k}$$

7. $\mathbf{j} \cdot (\mathbf{i} \times \mathbf{k}) = \mathbf{j} \cdot (-\mathbf{j}) = -1$ **9.** $(\mathbf{i} \times \mathbf{j}) \times \mathbf{k} = \mathbf{k} \times \mathbf{k} = 0$ **11.** $\mathbf{j} \cdot (\mathbf{k} \times \mathbf{i}) = \mathbf{j} \cdot (\mathbf{j}) = 1$

13. $(\mathbf{i} + 3\mathbf{j} - \mathbf{k}) \times (\mathbf{i} + \mathbf{k}) = \begin{vmatrix} \mathbf{i} & \mathbf{j} & \mathbf{k} \\ 1 & 3 & -1 \\ 2 & 0 & 1 \end{vmatrix} = [(3)(1) - (-1)(0)]\mathbf{i} - [(1)(1) - (-1)(1)]\mathbf{j} + [(1)0 - (3)(1)]\mathbf{k}$

$$= 3\mathbf{i} - 2\mathbf{j} - 3\mathbf{k}$$

15. $(\mathbf{i} + \mathbf{j} + \mathbf{k}) \times (2\mathbf{i} + \mathbf{k}) = \begin{vmatrix} \mathbf{i} & \mathbf{j} & \mathbf{k} \\ 1 & 1 & 1 \\ 2 & 0 & 1 \end{vmatrix} = [(1)(1) - (1)(0)]\mathbf{i} - [(1)(1) - (1)(2)]\mathbf{j} + [(1)(0) - (1)(2)]\mathbf{k}$

$$= \mathbf{i} + \mathbf{j} - 2\mathbf{k}$$

17. $[2\mathbf{i} + \mathbf{j}] \cdot [(\mathbf{i} - 3\mathbf{j} + \mathbf{k}) \times (4\mathbf{i} + \mathbf{k})] = \begin{vmatrix} 1 & -3 & 1 \\ 4 & 0 & 1 \\ 2 & 1 & 0 \end{vmatrix} =$

$$[(0)(0) - (1)(1)] - (-3)[(4)(0) - (1)(2)] + [(4)(1) - (0)(2)] = -3$$

19.

$$[(\mathbf{i} - \mathbf{j}) \times (\mathbf{j} - \mathbf{k})] \times [\mathbf{i} + 5\mathbf{k}] = \{[\mathbf{i} \times (\mathbf{j} - \mathbf{k})] - [\mathbf{j} \times (\mathbf{j} - \mathbf{k})]\} \times [\mathbf{i} + 5\mathbf{k}]$$

$$= [(\mathbf{k} + \mathbf{j}) - (-\mathbf{i})] \times [\mathbf{i} + 5\mathbf{k}]$$

$$= (\mathbf{i} + \mathbf{j} + \mathbf{k}) \times (\mathbf{i} + 5\mathbf{k})$$

$$= [(\mathbf{i} + \mathbf{j} + \mathbf{k}) \times \mathbf{i}] + [(\mathbf{i} + \mathbf{j} + \mathbf{k}) \times 5\mathbf{k}]$$

$$= (-\mathbf{k} + \mathbf{j}) + (-5\mathbf{j} + 5\mathbf{i})$$

$$= 5\mathbf{i} - 4\mathbf{j} - \mathbf{k}$$

21. $\mathbf{a} \times \mathbf{b} = \begin{vmatrix} 1 & -3 & 1 \\ 4 & 0 & 1 \\ 2 & 1 & 0 \end{vmatrix} = 3\mathbf{i} - 3\mathbf{j} - 6\mathbf{k}$

$\dfrac{\mathbf{a} \times \mathbf{b}}{\|\mathbf{a} \times \mathbf{b}\|} = \dfrac{1}{\sqrt{6}}\mathbf{i} - \dfrac{1}{\sqrt{6}}\mathbf{j} - \dfrac{2}{\sqrt{6}}\mathbf{k}; \quad \dfrac{\mathbf{b} \times \mathbf{a}}{\|\mathbf{b} \times \mathbf{a}\|} = -\dfrac{1}{\sqrt{6}}\mathbf{i} + \dfrac{1}{\sqrt{6}}\mathbf{j} + \dfrac{2}{\sqrt{6}}\mathbf{k}$

23. Set $\mathbf{a} = \overrightarrow{PQ} = -\mathbf{i} + 2\mathbf{k}$ and $\mathbf{b} = \overrightarrow{PR} = 2\mathbf{i} - \mathbf{k}$. Then

$$\mathbf{N} = \overrightarrow{PQ} \times \overrightarrow{PR} = \begin{vmatrix} \mathbf{i} & \mathbf{j} & \mathbf{k} \\ -1 & 0 & 2 \\ 2 & 0 & -1 \end{vmatrix} = 3\mathbf{j}$$

and $A = \frac{1}{2}\|\mathbf{a} \times \mathbf{b}\| = \frac{1}{2}\|3\mathbf{j}\| = \frac{3}{2}$.

25. Set $\mathbf{a} = \overrightarrow{PQ} = \mathbf{i} + \mathbf{j} - 3\mathbf{k}$ and $\mathbf{b} = \overrightarrow{PR} = -\mathbf{i} + 3\mathbf{j} - \mathbf{k}$. Then

$$\mathbf{N} = \overrightarrow{PQ} \times \overrightarrow{PR} = \begin{vmatrix} \mathbf{i} & \mathbf{j} & \mathbf{k} \\ 1 & 1 & -3 \\ -1 & 3 & -1 \end{vmatrix} = 8\mathbf{j} + 4\mathbf{j} + 4\mathbf{k}$$

and $A = \frac{1}{2}\|\mathbf{a} \times \mathbf{b}\| = \frac{1}{2}\|8\mathbf{i} + 4\mathbf{j} + 4\mathbf{k}\| = \frac{1}{2}\sqrt{8^2 + 4^2 + 4^2} = 2\sqrt{6}$.

27. $V = \left|[(\mathbf{i} + \mathbf{j}) \times (2\mathbf{i} - \mathbf{k})] \cdot (3\mathbf{j} + \mathbf{k})\right| = \left|(-\mathbf{i} + \mathbf{j} - 2\mathbf{k}) \cdot (3\mathbf{j} + \mathbf{k})\right| = 1$

29. $V = \overrightarrow{OP} \cdot \left(\overrightarrow{OQ} \times \overrightarrow{OR}\right) = \begin{vmatrix} 1 & 2 & 3 \\ 1 & 1 & 2 \\ 2 & 1 & 1 \end{vmatrix} = 2$

31. $(\mathbf{a} + \mathbf{b}) \times (\mathbf{a} - \mathbf{b}) = [\mathbf{a} \times (\mathbf{a} - \mathbf{b})] + [\mathbf{b} \times (\mathbf{a} - \mathbf{b})]$

$$= [\mathbf{a} \times (-\mathbf{b})] + [\mathbf{b} \times \mathbf{a}]$$

$$= -(\mathbf{a} \times \mathbf{b}) - (\mathbf{a} \times \mathbf{b}) = -2(\mathbf{a} \times \mathbf{b})$$

33. $\mathbf{a} \times \mathbf{i} = 0$, $\mathbf{a} \times \mathbf{j} = 0$ $\implies$ $\mathbf{a} \| \mathbf{i}$ and $\mathbf{a} \| \mathbf{j}$ $\implies$ $\mathbf{a} = 0$

35.
$$(\alpha \mathbf{a} + \beta \mathbf{b}) \times (\gamma \mathbf{a} + \delta \mathbf{b}) = (\alpha \mathbf{a} \times \delta \mathbf{b}) + (\beta \mathbf{b} \times \gamma \mathbf{a})$$

$$= \alpha \delta (\mathbf{a} \times \mathbf{b}) - \beta \gamma (\mathbf{a} \times \mathbf{b})$$

$$= (\alpha \delta - \beta \gamma)(\mathbf{a} \times \mathbf{b}) = \begin{vmatrix} \alpha & \beta \\ \gamma & \delta \end{vmatrix} (\mathbf{a} \times \mathbf{b})$$

37. $\mathbf{a} \cdot (\mathbf{b} \times \mathbf{c}) = (\mathbf{a} \times \mathbf{b}) \cdot \mathbf{c} = (\mathbf{c} \times \mathbf{a}) \cdot \mathbf{b} = (\mathbf{b} \times \mathbf{c}) \cdot \mathbf{a} = (\mathbf{a} \times -\mathbf{c}) \cdot \mathbf{b}$

$\mathbf{a} \cdot (\mathbf{c} \times \mathbf{b}) = \mathbf{c} \cdot (\mathbf{b} \times \mathbf{a}) = (-\mathbf{a} \times \mathbf{b}) \cdot \mathbf{c}$

39. $\mathbf{a} \times \mathbf{b}$ is perpendicular to the plane determined by $\mathbf{a}$ and $\mathbf{b}$;

$\mathbf{c}$ is in this plane iff $\mathbf{a} \times \mathbf{b} \cdot \mathbf{c} = 0$.

41. $\mathbf{a} \cdot \mathbf{b} = \mathbf{a} \cdot \mathbf{c}$ $\implies$ $\mathbf{a} \cdot (\mathbf{b} - \mathbf{c}) = 0$; $\mathbf{a}$ is perpendicular to $\mathbf{b} - \mathbf{c}$.

$\mathbf{a} \times \mathbf{b} = \mathbf{a} \times \mathbf{c}$ $\implies$ $\mathbf{a} \times (\mathbf{b} - \mathbf{c}) = 0$; $\mathbf{a}$ is parallel to $\mathbf{b} - \mathbf{c}$.

Since $\mathbf{a} \neq 0$ it follows that $\mathbf{b} - \mathbf{c} = 0$ or $\mathbf{b} = \mathbf{c}$.

43. $\mathbf{c} \times \mathbf{a} = (\mathbf{a} \times \mathbf{b}) \times \mathbf{a} = (\mathbf{a} \cdot \mathbf{a})\mathbf{b} - (\mathbf{a} \cdot \mathbf{b})\mathbf{a} = (\mathbf{a} \cdot \mathbf{a})\mathbf{b} = \|\mathbf{a}\|^2 \mathbf{b}$

$\quad\quad$ └─Exercise 42(a) $\quad\quad$ └─$\mathbf{a} \cdot \mathbf{b} = 0$

45. Expanding the determinant by the bottom row gives
$$\begin{vmatrix} a_1 & a_2 & a_3 \\ b_1 & b_2 & b_3 \\ c_1 & c_2 & c_3 \end{vmatrix} = c_1 \begin{vmatrix} a_2 & a_3 \\ b_2 & b_3 \end{vmatrix} - c_2 \begin{vmatrix} a_1 & a_3 \\ b_1 & b_3 \end{vmatrix} + c_3 \begin{vmatrix} a_1 & a_2 \\ b_1 & b_2 \end{vmatrix}$$

47. $\|\boldsymbol{\tau}\| = \|\mathbf{r}\| \cdot \|\mathbf{F}\| \sin\theta = (10)(20)\sin 50° = 153.21$ inch-lb $= 12.77$ ft-lb;

the bolt moves into the plane of the paper.

SECTION 12.5

1. P (when $t = 0$) and Q (when $t = -1$)

3. Take $\mathbf{r}_0 = \overrightarrow{OP} = 3\mathbf{i} + \mathbf{j}$ and $\mathbf{d} = \mathbf{k}$.   Then, $\mathbf{r}(t) = (3\mathbf{i} + \mathbf{j}) + t\mathbf{k}$.

5. Take $\mathbf{r}_0 = 0$ and $\mathbf{d} = \overrightarrow{OQ}$.   Then, $\mathbf{r}(t) = t(x_1\mathbf{i} + y_1\mathbf{j} + z_1\mathbf{k})$.

7. $\overrightarrow{PQ} = \mathbf{i} - \mathbf{j} + \mathbf{k}$ so direction numbers are $1, -1, 1$. Using P as a point on the line, we have

$$x(t) = 1 + t, \quad y(t) = -t, \quad z(t) = 3 + t.$$

9. The line is parallel to the y-axis so we can take $0, 1, 0$ as direction numbers. Therefore

$$x(t) = 2, \quad y(t) = -2 + t, \quad z(t) = 3.$$

11. Since the line $\quad 2(x+1) = 4(y-3) = z \quad$ can be written

$$\frac{x+1}{2} = \frac{y-3}{1} = \frac{z}{4},$$

it has direction numbers $2, 1, 4$. The line through $P(-1, 2, -3)$ with direction vector

$2\mathbf{i} + \mathbf{j} + 4\mathbf{k}$ can be parametrized

$$\mathbf{r}(t) = (-\mathbf{i} + 2\mathbf{j} - 3\mathbf{k}) + t(2\mathbf{i} + \mathbf{j} + 4\mathbf{k}).$$

13. We set $\quad \mathbf{r}_1(t) = \mathbf{r}_2(u) \quad$ and solve for t and u:

$$\mathbf{i} + t\mathbf{j} = \mathbf{j} + u(\mathbf{i} + \mathbf{j}),$$

$$(1 - u)\mathbf{i} + (-1 - u + t)\mathbf{j} = 0.$$

Thus,

$$1 - u = 0 \quad \text{and} \quad -1 - u + t = 0.$$

The equation gives $u = 1$, $t = 2$. The point of intersection is $P(1, 2, 0)$.

As direction vectors for the lines we can take $\mathbf{u} = \mathbf{j}$ and $\mathbf{v} = \mathbf{i} + \mathbf{j}$. Thus

$$\cos\theta = \frac{\mathbf{u} \cdot \mathbf{v}}{\|\mathbf{u}\| \, \|\mathbf{v}\|} = \frac{1}{(1)(\sqrt{2})} = \frac{1}{2}\sqrt{2}.$$

The angle of intersection is $\frac{1}{4}\pi$ radians.

15. We solve the system

$$3 + t = 1, \quad 1 - t = 4 + u, \quad 5 + 2t = 2 + u$$

for t and u to find that $t = -2$, $u = -1$. The point of intersection is $(1, 3, 1)$.

Since $\mathbf{i} - \mathbf{j} + 2\mathbf{k}$ is a direction vector for l_1 and $\mathbf{j} + \mathbf{k}$ is a direction vector for l_2,

$$\cos\theta = \frac{(\mathbf{i} - \mathbf{j} + 2\mathbf{k}) \cdot (\mathbf{j} + \mathbf{k})}{\sqrt{6}\sqrt{2}} = \frac{1}{2\sqrt{3}} = \frac{1}{6}\sqrt{3} \quad \text{and} \quad \theta \cong 1.28 \text{ radians.}$$

17. $\left(x_0 - \dfrac{d_1}{d_3} z_0, \; y_0 - \dfrac{d_2}{d_3} z_0, \; 0 \right)$ 19. The lines are parallel.

21. $\mathbf{r}(t) = (2\mathbf{i} + 7\mathbf{j} - \mathbf{k}) + t(2\mathbf{i} - 5\mathbf{j} + 4\mathbf{k}), \quad 0 \le t \le 1$

23. Set $\quad \mathbf{u} = \dfrac{\overrightarrow{PQ}}{\|\overrightarrow{PQ}\|} = \dfrac{-4\mathbf{i} + 2\mathbf{j} + 4\mathbf{k}}{\| -4\mathbf{i} + 2\mathbf{j} + 4\mathbf{k}\|} = -\dfrac{2}{3}\mathbf{i} + \dfrac{1}{3}\mathbf{j} + \dfrac{2}{3}\mathbf{k}.$

Then $\mathbf{r}(t) = (6\mathbf{i} - 5\mathbf{j} + \mathbf{k}) + t\mathbf{u}$ is $\overrightarrow{OP}$ at $t = 9$ and it is $\overrightarrow{OQ}$ at $t = 15$. (Check this.)

Answer: $\mathbf{u} = -\frac{2}{3}\mathbf{i} + \frac{1}{3}\mathbf{j} + \frac{2}{3}\mathbf{k}, \quad 9 \le t \le 15$.

25.   The given line, call it l, has direction vector $2\mathbf{i} - 4\mathbf{j} + 6\mathbf{k}$.

If $a\mathbf{i} + b\mathbf{j} + c\mathbf{k}$ is a direction vector for a line perpendicular to l, then

$$(2\mathbf{i} - 4\mathbf{j} + 6\mathbf{k}) \cdot (a\mathbf{i} + b\mathbf{j} + c\mathbf{k}) = 2a - 4b + 6c = 0.$$

The lines through $P(3, -1, 8)$ perpendicular to l can be parametrized

$$X(u) = 3 + au, \quad Y(u) = -1 + bu, \quad Z(u) = 8 + cu$$

with $2a - 4b + 6c = 0$.

27. $d(P, l) = \dfrac{\|(\mathbf{i} + 2\mathbf{k}) \times (2\mathbf{i} - \mathbf{j} + 2\mathbf{k})\|}{\|2\mathbf{i} - \mathbf{j} + 2\mathbf{k}\|} = 1$

29.   The line contains the point $P_0(1, 0, 2)$. Therefore

$$d(P, l) = \frac{\|(2\mathbf{j} + \mathbf{k}) \times (\mathbf{i} - 2\mathbf{j} + 3\mathbf{k})\|}{\|\mathbf{i} - 2\mathbf{j} + 3\mathbf{k}\|} = \sqrt{\frac{69}{14}} \cong 2.22$$

31.   The line contains the point $P_0(2, -1, 0)$. Therefore

$$d(P, l) = \frac{\|(\mathbf{i} - \mathbf{j} - \mathbf{k}) \times (\mathbf{i} + \mathbf{j})\|}{\|\mathbf{i} + \mathbf{j}\|} = \sqrt{3} \cong 1.73.$$

33.   (a)   The line passes through $P(1, 1, 1)$ with direction vector $\mathbf{i} + \mathbf{j}$. Therefore

$$d(0, l) = \frac{\|(\mathbf{i} + \mathbf{j} + \mathbf{k}) \times (\mathbf{i} + \mathbf{j})\|}{\|\mathbf{i} + \mathbf{j}\|} = 1.$$

(b)   The distance from the origin to the line segment is $\sqrt{3}$.

Solution. The line segment can be parametrized

$$\mathbf{r}(t) = \mathbf{i} + \mathbf{j} + \mathbf{k} + t(\mathbf{i} + \mathbf{j}), \quad t \in [0, 1].$$

This is the set of all points $P(1 + t, 1 + t, 1)$ with $t \in [0, 1]$.

The distance from the origin to such a point is

$$f(t) = \sqrt{2(1 + t^2) + 1}.$$

The minimum value of this function is $f(0) = \sqrt{3}$.

Explanation. The point on the line through P and Q closest to the origin is not on the line segment $\overline{PQ}$.

35.   We begin with $\mathbf{r}(t) = \mathbf{j} - 2\mathbf{k} + t(\mathbf{i} - \mathbf{j} + 3\mathbf{k})$. The scalar t_0 for which $\mathbf{r}(t_0) \perp l$ can be found by solving the equation

$$[\mathbf{j} - 2\mathbf{k} + t_0(\mathbf{i} - \mathbf{j} + 3\mathbf{k})] \cdot [\mathbf{i} - \mathbf{j} + 3\mathbf{k}] = 0.$$

This equation gives $-7 + 11t_0 = 0$ and thus $t_0 = 7/11$. Therefore

$$\mathbf{r}(t_0) = \mathbf{j} - 2\mathbf{k} + \tfrac{7}{11}(\mathbf{i} - \mathbf{j} + 3\mathbf{k}) = \tfrac{7}{11}\mathbf{i} + \tfrac{4}{11}\mathbf{j} - \tfrac{1}{11}\mathbf{k}.$$

The vectors of norm 1 parallel to $\mathbf{i} - \mathbf{j} + 3\mathbf{k}$ are

$$\pm \frac{1}{\sqrt{11}}(\mathbf{i} - \mathbf{j} + 3\mathbf{k}).$$

The standard parametrizations are

$$\mathbf{R}(t) = \frac{7}{11}\mathbf{i} + \frac{4}{11}\mathbf{j} - \frac{1}{11}\mathbf{k} \pm \frac{t}{\sqrt{11}}(\mathbf{i} - \mathbf{j} + 3\mathbf{k})$$

$$= \frac{1}{11}(7\mathbf{i} + 4\mathbf{j} - \mathbf{k}) \pm t\left[\frac{\sqrt{11}}{11}(\mathbf{i} - \mathbf{j} + 3\mathbf{k})\right].$$

37. $0 < t < s$

By similar triangles, if $0 < s < 1$, the tip of $\overrightarrow{OA} + s\overrightarrow{AB} + s\overrightarrow{BC}$ falls on $\overline{AC}$. If $0 < t < s$, then the tip of $\overrightarrow{OA} + s\overrightarrow{AB} + t\overrightarrow{BC}$ falls short of $\overline{AC}$ and stays within the triangle. Clearly all points in the interior of the triangle can be reached in this manner.

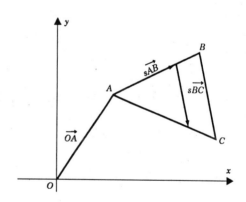

SECTION 12.6

1. Q

3.   Since $\mathbf{i} - 4\mathbf{j} + 3\mathbf{k}$ is normal to the plane, we have

$$(x - 2) - 4(y - 3) + 3(z - 4) = 0 \quad \text{and thus} \quad x - 4y + 3z - 2 = 0.$$

5.   The vector $3\mathbf{i} - 2\mathbf{j} + 5\mathbf{k}$ is normal to the given plane and thus to every parallel plane: the equation we want can be written

$$3(x - 2) - 2(y - 1) + 5(z - 1) = 0, \quad 3x - 2y + 5z - 9 = 0.$$

7.   The point $Q(0,0,-2)$ lies on the line l; and $\mathbf{d} = \mathbf{i} + \mathbf{j} + \mathbf{k}$ is a direction vector for l.

We want an equation for the plane which has the vector

$$N = \overrightarrow{PQ} \times \mathbf{d} = (\mathbf{i} + 3\mathbf{j} + 3\mathbf{k}) \times (\mathbf{i} + \mathbf{j} + \mathbf{k})$$

as a normal vector:

$$N = \begin{vmatrix} \mathbf{i} & \mathbf{j} & \mathbf{k} \\ 1 & 3 & 3 \\ 1 & 1 & 1 \end{vmatrix} = 2\mathbf{j} - 2\mathbf{k}$$

An equation for the plane is: $2(y-3) - 2(z-1) = 0$ or $y - z - 2 = 0$

9. $\overrightarrow{OP_0} = x_0\mathbf{i} + y_0\mathbf{j} + z_0\mathbf{k}$ An equation for the plane is:

$$x_0(x - x_0) + y_0(y - y_0) + z_0(z - z_0)$$

11.   The vector $N = 2\mathbf{i} - \mathbf{j} + 5\mathbf{k}$ is normal to the plane $2x - y + 5z - 10 = 0$. The unit normals are:

$$\frac{N}{\|N\|} = \frac{1}{\sqrt{30}}(2\mathbf{i} - \mathbf{j} + 5\mathbf{k}) \quad \text{and} \quad -\frac{N}{\|N\|} = -\frac{1}{\sqrt{30}}(2\mathbf{i} - \mathbf{j} + 5\mathbf{k})$$

13.   Intercept form: $\dfrac{x}{15} + \dfrac{y}{12} - \dfrac{z}{10} = 1$ x-intercept: $(15, 0, 0)$

y-intercept: $(0, 12, 0)$

z-intercept: $(0, 0, 10)$

15. $\mathbf{u}_{N_1} = \dfrac{\sqrt{38}}{38}(5\mathbf{i} - 3\mathbf{j} + 2\mathbf{k}), \quad \mathbf{u}_{N_2} = \dfrac{\sqrt{14}}{14}(\mathbf{i} + 3\mathbf{j} + 2\mathbf{k}), \quad \cos\theta = |\mathbf{u}_{N_1} \cdot \mathbf{u}_{N_2}| = 0.$

Therefore $\theta = \pi/2$ radians.

17. $\mathbf{u}_{N_1} = \dfrac{\sqrt{3}}{3}(\mathbf{i} - \mathbf{j} + \mathbf{k}), \quad \mathbf{u}_{N_2} = \dfrac{\sqrt{14}}{14}(2\mathbf{i} + \mathbf{j} + 3\mathbf{k}), \quad \cos\theta = |\mathbf{u}_{N_1} \cdot \mathbf{u}_{N_2}| = \dfrac{2}{21}\sqrt{42} \cong 0.617.$

Therefore $\theta \cong 0.91$ radians.

19.   coplanar since $0(4\mathbf{j} - \mathbf{k}) + 0(3\mathbf{i} + \mathbf{j} + 2\mathbf{k}) + 1(0) = 0$

21.   We need to determine whether there exist scalars s, t, u not all zero such that

$$s(\mathbf{i} + \mathbf{j} + \mathbf{k}) + t(2\mathbf{i} - \mathbf{j}) + u(3\mathbf{i} - \mathbf{j} - \mathbf{k}) = 0$$

$$(s + 2t + 3u)\mathbf{i} + (s - t - u)\mathbf{j} + (s - u)\mathbf{k} = \mathbf{0}.$$

The only solution of the system

$$s + 2t + 3u = 0, \quad s - t - u = 0, \quad s - u = 0$$

is $s = t = u = 0$. Thus, the vectors are not coplanar.

23. By (12.6.7), $\quad d(P,p) = \dfrac{|2(2) + 4(-1) - (3) + 1|}{\sqrt{4 + 16 + 1}} = \dfrac{2}{\sqrt{21}} = \dfrac{2}{21}\sqrt{21}.$

25. By (12.6.7), $\quad d(P,p) = \dfrac{|(-3)(1) + 0(-3) + 4(5) + 5|}{\sqrt{9 + 16}} = \dfrac{22}{5}.$

27. $\overrightarrow{P_1P} = (x-1)\mathbf{i} + y\mathbf{j} + (z-1)\mathbf{k}, \quad \overrightarrow{P_1P_2} = \mathbf{i} + \mathbf{j} - \mathbf{k}, \quad \overrightarrow{P_1P_3} = \mathbf{j}.$

Therefore

$$(\overrightarrow{P_1P_2} \times \overrightarrow{P_1P_3}) = (\mathbf{i} + \mathbf{j} - \mathbf{k}) \times \mathbf{j} = \mathbf{i} + \mathbf{k}$$

and

$$\overrightarrow{P_1P} \cdot (\overrightarrow{P_1P_2} \times \overrightarrow{P_1P_3}) = [(x-1)\mathbf{i} + y\mathbf{j} + (z-1)\mathbf{k}] \cdot [\mathbf{i} + \mathbf{k}] = x - 1 + z - 1.$$

An equation for the plane can be written $\quad x + z = 2.$

29. $\overrightarrow{P_1P} = (x-3)\mathbf{i} + (y+4)\mathbf{j} + (z-1)\mathbf{k}, \quad \overrightarrow{P_1P_2} = 6\mathbf{j}, \quad \overrightarrow{P_1P_3} = -4\mathbf{i} + 5\mathbf{j} - 3\mathbf{k}.$

Therefore

$$(\overrightarrow{P_1P_2} \times \overrightarrow{P_1P_3}) = 6\mathbf{j} \times (-4\mathbf{i} + 5\mathbf{j} - 3\mathbf{k}) = -18\mathbf{i} + 24\mathbf{k}$$

and $\quad \overrightarrow{P_1P} \cdot (\overrightarrow{P_1P_2} \times \overrightarrow{P_1P_3}) = [(x-3)\mathbf{i} + (y+4)\mathbf{j} + (z-1)\mathbf{k}] \cdot [-18\mathbf{i} + 24\mathbf{k}]$

$$= -18(x-3) + 24(z-1)$$

An equation for the plane can be written $\quad -18(x-3) + 24(z-1) = 0 \quad$ or $\quad 3x - 4z - 5 = 0.$

31. The line passes through the point $P_0\,(x_0, y_0, z_0)$ with direction numbers: A, B, C.

Equations for the line written in symmetric form are:

$$\frac{x - x_0}{A} = \frac{y - y_0}{B} = \frac{z - z_0)}{C}, \quad \text{provided} \quad A \neq 0,\ B \neq 0,\ C \neq 0.$$

33. $\dfrac{x - x_0}{d_1} = \dfrac{y - y_0}{d_2}, \qquad \dfrac{y - y_0}{d_2} = \dfrac{z - z_0}{d_3}$

35. Following the hint we take $x = 0$ and find that $P_0(0,0,0)$ lies on the line of intersection. As normals to the plane we use

$$\mathbf{N}_1 = \mathbf{i} + 2\mathbf{j} + 3\mathbf{k} \quad \text{and} \quad \mathbf{N}_2 = -3\mathbf{i} + 4\mathbf{j} + \mathbf{k}.$$

Note that

$$\mathbf{N_1} \times \mathbf{N_2} = (\mathbf{i} + 2\mathbf{j} + 3\mathbf{k}) \times (-3\mathbf{i} + 4\mathbf{j} + \mathbf{k}) = -10\mathbf{i} - 10\mathbf{j} + 10\mathbf{k}.$$

We take $-\frac{1}{10}(\mathbf{N_1} \times \mathbf{N_2}) = \mathbf{i} + \mathbf{j} - \mathbf{k}$ as a direction vector for the line through $P_0(0,0,0)$. Then

$$x(t) = t, \quad y(t) = t, \quad z(t) = -t.$$

37. Straightforward computations give us

$$l: x(t) = 1 - 3t, \quad y(t) = -1 + 4t, \quad z(t) = 2 - t$$

and

$$p: x + 4y - z = 6.$$

Substitution of the scalar parametric equations for l in the equation for p gives

$$(1 - 3t) + 4(-1 + 4t) - (2 - t) = 6 \quad \text{and thus} \quad t = 11/14.$$

Using $t = 11/14$, we get $\quad x = -19/14, \quad y = 15/7, \quad z = 17/14$.

39. Let $\quad \mathbf{N} = A\mathbf{i} + B\mathbf{j} + C\mathbf{k} \quad$ be normal to the plane. Then

$$\mathbf{N} \cdot \mathbf{d} = (\mathbf{i} + B\mathbf{j} + C\mathbf{k}) \cdot (\mathbf{i} + 2\mathbf{j} + 4\mathbf{k}) = 1 + 2B + 4C = 0$$

and

$$\mathbf{N} \cdot \mathbf{d} = (\mathbf{i} + B\mathbf{j} + C\mathbf{k}) \cdot (-\mathbf{i} - \mathbf{j} + 3\mathbf{k}) = -1 - B + 3C = 0.$$

This gives $\quad B = -7/10 \quad$ and $\quad C = 1/10$. The equation for the plane can be written

$$1(x - 0) - \tfrac{7}{10}(y - 0) + \tfrac{1}{10}(z - 0) = 0, \quad \text{which simplifies to} \quad 10x - 7y + z = 0.$$

41. $\mathbf{N} + \overrightarrow{PQ}$ and $\mathbf{N} - \overrightarrow{PQ}$ are the diagonals of a rectangle with sides $\mathbf{N}$ and $\overrightarrow{PQ}$. Since the diagonals are perpendicular, the rectangle is a square; that is $\|\mathbf{N}\| = \|\overrightarrow{PQ}\|$. Thus, the points Q form a circle centered at P with radius $\|\mathbf{N}\|$.

43. If $\alpha > 0$, then P_1 lies on the same side of the plane as the tip of $\mathbf{N}$; if $\alpha < 0$, then P_1 and the tip of $\mathbf{N}$ lie on opposite sides of the plane.

To see this, suppose that the tip of $\mathbf{N}$ is at $P_0(x_0, y_0, z_0)$. Then

$$\mathbf{N} \cdot \overrightarrow{P_0 P_1} = A(x_1 - x_0) + B(y_1 - y_0) + C(z_1 - z_0) = Ax_1 + By_1 + Cz_1 + D = \alpha.$$

If $\alpha > 0$, $0 \leq \measuredangle \left(\mathbf{N}, \overrightarrow{P_0 P_1} \right) < \pi/2$; if $\alpha < 0$, then $\pi/2 < \measuredangle \left(\mathbf{N}, \overrightarrow{P_0 P_1} \right) < \pi$. Since $\mathbf{N}$ is perpendicular to the plane, the result follows.

45. $\quad \mathbf{a} \cdot (\mathbf{b} \times \mathbf{c}) = 0$

47.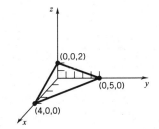

49.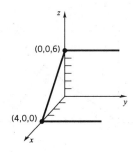

51. $\dfrac{x}{2} + \dfrac{y}{5} + \dfrac{z}{4} = 1$

$10x + 4y + 5z = 20$

53. $\dfrac{x}{3} + \dfrac{y}{5} = 1$

$5x + 3y = 15$

PROJECTS AND EXPLORATIONS

12.1. **(a)**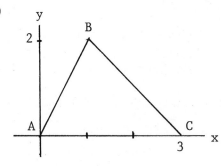

The median vectors are: $(2,1)$, $(-2.5,1)$, $(0.5,-2)$.

Their lengths are: $\sqrt{5} \cong 2.236$, $\sqrt{7.25} \cong 2.693$, $\sqrt{4.25} \cong 2.062$.

The triangle inequality is clearly satisfied by these values.

(b) Perimeter of T : 8.06450;   perimeter of M : 12.09674

Area of T : 3;   area of M : 6.75

(c) Angles of T : $A \cong 1.10715 \cong 63.43°$, $B \cong 1.24905 \cong 71.57°$, $C \cong 0.78540 = 45°$

Angles of M : $\alpha \cong 0.78540 = 45°$, $\beta = 1.10715 \cong 63.43°$, $\gamma \cong 1.24905 \cong 71.57°$

(d) Triangles T and M are similar.

(e)

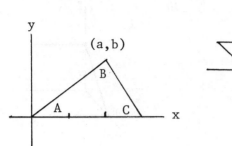 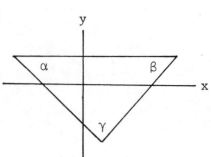

The median vectors are: $\left(\dfrac{3+a}{2}, \dfrac{b}{2}\right),$ $\left(\dfrac{3}{2}-a, -b\right),$ $\left(\dfrac{a}{2}-3, \dfrac{b}{2}\right)$

Their lengths are: $\frac{1}{2}\sqrt{(3+a)^2+b^2}$, $\frac{1}{2}\sqrt{(3-2a)^2+4b^2}$, $\frac{1}{2}\sqrt{(a-6)^2+b^2}$;

the triangle inequality is satisfied.

Perimeter of T : $3 + \sqrt{a^2+b^2} + \sqrt{(3-a)^2+b^2}$,

perimeter of M : $4.5 + \frac{3}{2}\sqrt{a^2+b^2} + \frac{3}{2}\sqrt{(3-a)^2+b^2} = \frac{3}{2}$ (perimeter of T).

Area of T : $\frac{3}{2}b$:   area of M : $\frac{27}{8}b = \frac{9}{4}$ (area of T)

Angles: $\tan A = \dfrac{b}{a} = \tan \beta,$ $\tan C = \dfrac{b}{3-a} = \tan \alpha$ $\Rightarrow$ $B = \gamma$;

Triangles T and M are similar.

12.3.   (a) Given the points $A\,(0,0,0)$, $B\,(4,0,0)$, $C\,(3,2,1)$. Let $0 \le x \le 1$, $0 \le y \le 1$. Then

$$P\,[4x(1-y)+3y,\ 2y,\ y] = (1-x)A + x[(1-y)B + yC]$$

Now, $Q = (1-y)B + yC$ lies on the line segment joining B and C and $P = (1-x)A + xQ$ lies on the segment joining P and Q. It follows that P is either inside or on the boundary of R.

(b) $f(x,y) = \sqrt{[4x(1-y)+3y]^2 + 5y^2} + \sqrt{[4 - 4x(1-y) - 3y]^2 + 5y^2} +$

$$\sqrt{[3 - 4x(1-y) - 3y]^2 + 5(1-y)^2}.$$

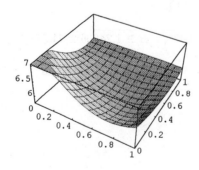

(c) Absolute minimum occurs at $x = 0.65193$, $y = 0.45472$;

$f(0.65193, 0.45472) \cong 5.78722$

Absolute maximum occurs at vertex A; $f(A) \cong 7.74166$.

(d) At the minimum, P is the point $(2.78600, 0.90945, 0.45472)$;

The angles between the vectors are $120°$.

CHAPTER 13

SECTION 13.1

1. $\mathbf{f}'(t) = 2\mathbf{i} - \mathbf{j} + 3\mathbf{k}$

3. $\mathbf{f}'(t) = -\dfrac{1}{2\sqrt{1-t}}\mathbf{i} + \dfrac{1}{2\sqrt{1+t}}\mathbf{j} + \dfrac{1}{(1-t)^2}\mathbf{k}$

5. $\mathbf{f}'(t) = \cos t\,\mathbf{i} - \sin t\,\mathbf{j} + \sec^2 t\,\mathbf{k}$

7. $\mathbf{f}'(t) = \dfrac{-1}{1-t}\mathbf{i} - \sin t\,\mathbf{j} + 2t\,\mathbf{k}$

9. $\mathbf{f}'(t) = 4\mathbf{i} + 6t^2\mathbf{j} + (2t+2)\mathbf{k};\qquad \mathbf{f}''(t) = 12t\,\mathbf{j} + 2\mathbf{k}$

11. $\mathbf{f}'(t) = (e^t\cos t - e^t\sin t)\mathbf{i} + (e^t\sin t + e^t\cos t)\mathbf{j} + 4\mathbf{k};\quad \mathbf{f}''(t) = -2e^t\sin t\,\mathbf{i} + 2e^t\cos t\,\mathbf{j}$

13. $\displaystyle\int_1^2 (\mathbf{i} + 2t\,\mathbf{j})\,dt = \left[t\,\mathbf{i} + t^2\mathbf{j}\right]_1^2 = \mathbf{i} + 3\mathbf{j}$

15. $\displaystyle\int_0^1 (e^t\mathbf{i} + e^{-t}\mathbf{k})\,dt = \left[e^t\mathbf{i} - e^{-t}\mathbf{k}\right]_0^1 = (e-1)\mathbf{i} + \left(1 - \dfrac{1}{e}\right)\mathbf{k}$

17. $\displaystyle\int_0^1 \left(\dfrac{1}{1+t^2}\mathbf{i} + \sec^2 t\,\mathbf{j}\right)dt = \left[\tan^{-1} t\,\mathbf{i} + \tan t\,\mathbf{j}\right]_0^1 = \dfrac{\pi}{4}\mathbf{i} + \tan(1)\mathbf{j}$

19. $\displaystyle\lim_{t\to 0}\mathbf{f}(t) = \left(\lim_{t\to 0}\dfrac{\sin t}{2t}\right)\mathbf{i} + \left(\lim_{t\to 0}e^{2t}\right)\mathbf{j} + \left(\lim_{t\to 0}\dfrac{t^2}{e^t}\right)\mathbf{k} = \dfrac{1}{2}\mathbf{i} + \mathbf{j}$

21. $\displaystyle\lim_{t\to 0}\mathbf{f}(t) = \left(\lim_{t\to 0}t^2\right)\mathbf{i} + \left(\lim_{t\to 0}\dfrac{1-\cos t}{3t}\right)\mathbf{j} + \left(\lim_{t\to 0}\dfrac{t}{t+1}\right)\mathbf{k} = 0\mathbf{i} + \dfrac{1}{3}\left(\lim_{t\to 0}\dfrac{1-\cos t}{t}\right)\mathbf{j} + 0\mathbf{k} = \mathbf{0}$

23. The limit does not exist since $\displaystyle\lim_{t\to 0}\dfrac{1}{t}$ does not exist.

25.    27. 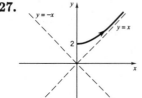   29. 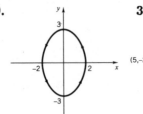   31.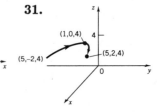

33. 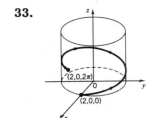   35. 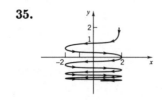   37.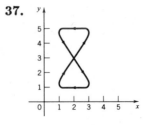

39.   (a) $\mathbf{f}(t) = 3\cos t\,\mathbf{i} + 2\sin t\,\mathbf{j}$ (b) $\mathbf{f}(t) = 3\cos t\,\mathbf{i} - 2\sin t\,\mathbf{j}$

41.   (a) $\mathbf{f}(t) = t\mathbf{i} + t^2\mathbf{j}$ (b) $\mathbf{f}(t) = -t\mathbf{i} + t^2\mathbf{j}$

43. $\mathbf{f}(t) = (1 + 2t)\mathbf{i} + (4 + 5t)\mathbf{j} + (-2 + 8t)\mathbf{k}, \quad 0 \le t \le 1$

45. $\mathbf{f}'(t_0) = \mathbf{i} + m\mathbf{j}$,

$$\int_a^b \mathbf{f}(t)\,dt = \left[\tfrac{1}{2}t^2\mathbf{i}\right]_a^b + \left[\int_a^b f(t)\,dt\right]\mathbf{j} = \tfrac{1}{2}\left(b^2 - a^2\right)\mathbf{i} + A\mathbf{j},$$

$$\int_a^b \mathbf{f}'(t)\,dt = [t\mathbf{i} + f(t)\mathbf{j}]_a^b = (b - a)\mathbf{i} + (d - c)\mathbf{j}$$

47.
$$\mathbf{f}'(t) = \mathbf{i} + t^2\mathbf{j}$$

$$\mathbf{f}(t) = (t + C_1)\mathbf{i} + \left(\tfrac{1}{3}t^3 + C_2\right)\mathbf{j} + C_3\mathbf{k}$$

$$\mathbf{f}(0) = \mathbf{j} - \mathbf{k} \implies C_1 = 0, \quad C_2 = 1, \quad C_3 = -1$$

$$\mathbf{f}(t) = t\mathbf{i} + \left(\tfrac{1}{3}t^3 + 1\right)\mathbf{j} - \mathbf{k}$$

49. $\mathbf{f}'(t) = \alpha\,\mathbf{f}(t) \implies \mathbf{f}(t) = e^{\alpha t}\,\mathbf{f}(0) = e^{\alpha t}\mathbf{c}$

51.   (a)   If $\mathbf{f}'(t) = \mathbf{0}$ on an interval, then the derivative of each component is 0 on that interval, each component is constant on that interval, and therefore $\mathbf{f}$ itself is constant on that interval.

      (b)   Set $\mathbf{h}(t) = \mathbf{f}(t) - \mathbf{g}(t)$ and apply part (a).

53.   If $\mathbf{f}$ is differentiable at t, then each component is differentiable at t, each component is continuous at t, and therefore $\mathbf{f}$ is continuous at t.

55.   no; as a counter-example, set $\mathbf{f}(t) = \mathbf{i} = \mathbf{g}(t)$.

57.   Suppose $\mathbf{f}(t) = f_1(t)\mathbf{i} + f_2(t)\mathbf{j} + f_3(t)\mathbf{k}$.   Then $\|\mathbf{f}(t)\| = \sqrt{f_1^2(t) + f_2^2(t) + f_3^2(t)}$ and

$$\frac{d}{dt}\left(\|\mathbf{f}\|\right) = \frac{1}{2}\left[f_1^2 + f_2^2 + f_3^2\right]^{-1/2}\left(2f_1 \cdot f_1' + 2f_2 \cdot f_2' + 2f_3 \cdot f_3'\right) = \frac{\mathbf{f}(t) \cdot \mathbf{f}'(t)}{\|\mathbf{f}(t)\|}$$

SECTION 13.2

1. $\mathbf{f}'(t) = \mathbf{b}, \quad \mathbf{f}''(t) = \mathbf{0}$

3. $\mathbf{f}'(t) = 2e^{2t}\,\mathbf{i} - \cos t\,\mathbf{j}, \quad \mathbf{f}''(t) = 4e^{2t}\,\mathbf{i} + \sin t\,\mathbf{j}$

5. $\mathbf{f}'(t) = [(t^2\mathbf{i} - 2t\mathbf{j}) \cdot (\mathbf{i} + 3t^2\mathbf{j}) + (2t\mathbf{i} - 2\mathbf{j}) \cdot (t\mathbf{i} + t^3\mathbf{j})]\mathbf{j} = [3t^2 - 8t^3]\mathbf{j}$

      $\mathbf{f}''(t) = (6t - 24t^2)\mathbf{j}$

7.
$$\mathbf{f}'(t) = \left[(e^t\mathbf{i} + t\mathbf{k}) \times \frac{d}{dt}(t\mathbf{j} + e^{-t}\mathbf{k})\right] + \left[\frac{d}{dt}(e^t\mathbf{i} + t\mathbf{k}) \times (t\mathbf{j} + e^{-t}\mathbf{k})\right]$$

$$= \left[(e^t\mathbf{i} + t\mathbf{k}) \times (\mathbf{j} - e^{-t}\mathbf{k})\right] + \left[(e^t\mathbf{i} + \mathbf{k}) \times (t\mathbf{j} + e^{-t}\mathbf{k})\right]$$

$$= (-t\mathbf{i} + \mathbf{j} + e^t\mathbf{k}) + (-t\mathbf{i} - \mathbf{j} + te^t\mathbf{k})$$

$$= -2t\mathbf{i} + e^t(t+1)\mathbf{k}$$

$$\mathbf{f}''(t) = -2\mathbf{i} + e^t(t+2)\mathbf{k}$$

9. $\mathbf{f}'(t) = (\cos t\,\mathbf{i} + \sin t\,\mathbf{j} + \mathbf{k}) \times (2\sin 2t\,\mathbf{i} - 2\cos 2t\,\mathbf{j} + \mathbf{k}) + (-\sin t\,\mathbf{i} + \cos t\,\mathbf{j}) \times (\sin 2t\,\mathbf{i} + \cos 2t\,\mathbf{j} + t\,\mathbf{k})$

$\quad = (\sin t + t\cos t + 2\sin 2t)\mathbf{i} + (2\cos 2t - \cos t + t\sin t)\mathbf{j} - 3\sin 3t\,\mathbf{k}$

$\mathbf{f}''(t) = (2\cos t - t\sin t + 4\cos t)\mathbf{i} + (-4\sin 2t + 2\sin t + t\cos t)\mathbf{j} - 9\cos 3t\,\mathbf{k}$

11. $\mathbf{f}'(t) = \dfrac{1}{2}\sqrt{t}\,\mathbf{g}'(\sqrt{t}) + \mathbf{g}(\sqrt{t}), \quad \mathbf{f}''(t) = \dfrac{1}{4}\mathbf{g}''(\sqrt{t}) + \dfrac{3}{4\sqrt{t}}\mathbf{g}'(\sqrt{t})$

13. $-\sin t\,e^{\cos t}\,\mathbf{i} + \cos t\,e^{\sin t}\,\mathbf{j}$

15. $(e^t\mathbf{i} + e^{-t}\mathbf{j}) \cdot (e^t\mathbf{i} - e^{-t}\mathbf{j}) = e^{2t} - e^{-2t}; \quad$ therefore

$$\frac{d^2}{dt^2}\left[(e^t\mathbf{i} + e^{-t}\mathbf{j}) \cdot (e^t\mathbf{i} - e^{-t}\mathbf{j})\right] = \frac{d^2}{dt^2}\left[e^{2t} - e^{-2t}\right] = \frac{d}{dt}\left[2e^{2t} + 2e^{-2t}\right] = 4e^{2t} - 4e^{-2t}$$

17. $\dfrac{d}{dt}\left[(\mathbf{a} + t\mathbf{b}) \times (\mathbf{c} + t\mathbf{d})\right] = \left[(\mathbf{a} + t\mathbf{b}) \times \mathbf{d}\right] + \left[\mathbf{b} \times (\mathbf{c} + t\mathbf{d})\right] = (\mathbf{a} \times \mathbf{d}) + (\mathbf{b} \times \mathbf{c}) + 2t\,(\mathbf{b} \times \mathbf{d})$

19. $\dfrac{d}{dt}\left[(\mathbf{a} + t\mathbf{b}) \cdot (\mathbf{c} + t\mathbf{d})\right] = \left[(\mathbf{a} + t\mathbf{b}) \cdot \mathbf{d}\right] + \left[\mathbf{b} \cdot (\mathbf{c} + t\mathbf{d})\right] = (\mathbf{a} \cdot \mathbf{d}) + (\mathbf{b} \cdot \mathbf{c}) + 2t\,(\mathbf{b} \cdot \mathbf{d})$

21. $\mathbf{r}(t) = \mathbf{a} + t\mathbf{b}$ **23.** $\mathbf{r}(t) = \frac{1}{2}t^2\mathbf{a} + \frac{1}{6}t^3\mathbf{b} + t\mathbf{c} + \mathbf{d}$

25. $\mathbf{r}(t) = \sin t\,\mathbf{i} + \cos t\,\mathbf{j}, \quad \mathbf{r}'(t) = \cos t\,\mathbf{i} - \sin t\,\mathbf{j}, \quad \mathbf{r}''(t) = -\sin t\,\mathbf{i} - \cos t\,\mathbf{j} = -\mathbf{r}(t).$

Thus $\mathbf{r}(t)$ and $\mathbf{r}''(t)$ are parallel, and they always point in opposite directions.

27.
$$\mathbf{r}(t) \cdot \mathbf{r}'(t) = (\cos t\,\mathbf{i} + \sin t\,\mathbf{j}) \cdot (-\sin t\,\mathbf{i} + \cos t\,\mathbf{j}) = 0$$

$$\mathbf{r}(t) \times \mathbf{r}'(t) = (\cos t\,\mathbf{i} + \sin t\,\mathbf{j}) \times (-\sin t\,\mathbf{i} + \cos t\,\mathbf{j})$$

$$= \cos^2 t\,\mathbf{k} + \sin^2 t\,\mathbf{k} = (\cos^2 t + \sin^2 t)\,\mathbf{k} = \mathbf{k}$$

29. $\dfrac{d}{dt}\left[\mathbf{f}(t) \times \mathbf{f}'(t)\right] = \left[\mathbf{f}(t) \times \mathbf{f}''(t)\right] + \underbrace{\left[\mathbf{f}'(t) \times \mathbf{f}'(t)\right]}_{0} = \mathbf{f}(t) \times \mathbf{f}''(t).$

31. $[\mathbf{f} \cdot \mathbf{g} \times \mathbf{h}]' = \mathbf{f}' \cdot (\mathbf{g} \times \mathbf{h}) + \mathbf{f} \cdot (\mathbf{g} \times \mathbf{h})' = \mathbf{f}' \cdot (\mathbf{g} \times \mathbf{h}) + \mathbf{f} \cdot [\mathbf{g}' \times \mathbf{h} + \mathbf{g} \times \mathbf{h}']$

and the result follows.

33. $\|\mathbf{r}(t)\|$ is constant $\iff$ $\|\mathbf{r}(t)\|^2 = \mathbf{r}(t) \cdot \mathbf{r}(t)$ is constant

$\iff \dfrac{d}{dt}[\mathbf{r}(t) \cdot \mathbf{r}(t)] = 2[\mathbf{r}(t) \cdot \mathbf{r}'(t)] = 0$ identically

$\iff \mathbf{r}(t) \cdot \mathbf{r}'(t) = 0$ identically

35. Write

$$\frac{[\mathbf{f}(t+h) \times \mathbf{g}(t+h)] - [\mathbf{f}(t) \times \mathbf{g}(t)]}{h}$$

as

$$\left(\mathbf{f}(t+h) \times \left[\frac{\mathbf{g}(t+h) - \mathbf{g}(t)}{h}\right]\right) + \left(\left[\frac{\mathbf{f}(t+h) - \mathbf{f}(t)}{h}\right] \times \mathbf{g}(t)\right)$$

and take the limit as $h \to 0$. (Appeal to Theorem 13.1.3.)

SECTION 13.3

1. $\mathbf{r}'(t) = -\pi \sin \pi t\, \mathbf{i} + \pi \cos \pi t\, \mathbf{j} + \mathbf{k}, \quad \mathbf{r}'(2) = \pi \mathbf{j} + \mathbf{k}$

$\mathbf{R}(u) = (\mathbf{i} + 2\mathbf{k}) + u(\pi \mathbf{j} + \mathbf{k})$

3. $\mathbf{r}'(t) = \mathbf{b} + 2t\mathbf{c}, \quad \mathbf{r}'(-1) = \mathbf{b} - 2\mathbf{c}, \quad \mathbf{R}(u) = (\mathbf{a} - \mathbf{b} + \mathbf{c}) + u(\mathbf{b} - 2\mathbf{c})$

5. $\mathbf{r}'(t) = 4t\mathbf{i} - \mathbf{j} + 4t\mathbf{k}, \quad P$ is tip of $\mathbf{r}(1), \quad \mathbf{r}'(1) = 4\mathbf{i} - \mathbf{j} + 4\mathbf{k}$

$\mathbf{R}(u) = (2\mathbf{i} + 5\mathbf{k}) + u(4\mathbf{i} - \mathbf{j} + 4\mathbf{k})$

7. $\mathbf{r}'(t) = -2 \sin t\, \mathbf{i} + 3 \cos t\, \mathbf{j} + \mathbf{k}, \quad \mathbf{r}'(\pi/4) = -\sqrt{2}\,\mathbf{i} + \frac{3}{2}\sqrt{2}\,\mathbf{j} + \mathbf{k}$

$\mathbf{R}(u) = \left(\sqrt{2}\,\mathbf{i} + \frac{3}{2}\sqrt{2}\,\mathbf{j} + \frac{\pi}{4}\mathbf{k}\right) + u\left(-\sqrt{2}\,\mathbf{i} + \frac{3}{2}\sqrt{2}\,\mathbf{j} + \mathbf{k}\right)$

9. The scalar components $x(t) = at$ and $y(t) = bt^2$ satisfy the equation

$$a^2 y(t) = a^2(bt^2) = b(a^2 t^2) = b[x(t)]^2$$

and generate the parabola $a^2 y = bx^2$.

11. $\mathbf{r}(t) = t\mathbf{i} + (1 + t^2)\mathbf{j}, \quad \mathbf{r}'(t) = \mathbf{i} + 2t\mathbf{j}$

(a) $\mathbf{r}(t) \perp \mathbf{r}'(t) \implies \mathbf{r}(t) \cdot \mathbf{r}'(t) = [t\mathbf{i} + (1 + t^2)\mathbf{j}] \cdot (\mathbf{i} + 2t\mathbf{j})$

$= t(2t^2 + 3) = 0 \implies t = 0$

$\mathbf{r}(t)$ and $\mathbf{r}'(t)$ are perpendicular at $(0, 1)$.

(b) and (c) $\mathbf{r}(t) = \alpha\,\mathbf{r}'(t)$ with $\alpha \neq 0$ $\implies$ $t = \alpha$ and $1 + t^2 = 2t\alpha$ $\implies$ $t = \pm 1$.

If $\alpha > 0$, then $t = 1$. $\mathbf{r}(t)$ and $\mathbf{r}'(t)$ have the same direction at $(1, 2)$.

If $\alpha < 0$, then $t = -1$. $\mathbf{r}(t)$ and $\mathbf{r}'(t)$ have opposite directions at $(-1, 2)$.

13. The tangent line at $t = t_0$ has the form $\mathbf{R}(u) = \mathbf{r}(t_0) + u\,\mathbf{r}'(t_0)$. If $\mathbf{r}'(t_0) = \alpha\,\mathbf{r}(t_0)$, then

$$\mathbf{R}(u) = \mathbf{r}(t_0) + u\,\alpha\mathbf{r}(t_0) = (1 + u\alpha)\,\mathbf{r}(t_0).$$

The tangent line passes through the origin at $u = -1/\alpha$.

15. $\mathbf{r}_1(t)$ passes through $P(0, 0, 0)$ at $t = 0$; $\mathbf{r}_2(u)$ passes through $P(0, 0, 0)$ at $u = -1$.

$$\mathbf{r}_1'(t) = e^t\mathbf{i} + 2\,\cos t\,\mathbf{j} + \frac{1}{t+1}\,\mathbf{k}; \quad \mathbf{r}_1'(0) = \mathbf{i} + 2\mathbf{j} + \mathbf{k}$$

$$\mathbf{r}_2'(u) = \mathbf{i} + 2u\mathbf{j} + 3u^2\,\mathbf{k}; \quad \mathbf{r}_2'(-1) = \mathbf{i} - 2\mathbf{j} + 3\mathbf{k}$$

$$\cos\theta = \frac{\mathbf{r}_1'(0)\,\cdot\,\mathbf{r}_2'(1)}{\|\mathbf{r}_1'(0)\|\,\|\mathbf{r}_2'(1)\|} = 0; \quad \theta = \frac{\pi}{2} \cong 1.57, \text{ or } 90°.$$

17. $\mathbf{r}_1(t) = \mathbf{r}_2(u)$ implies

$$\left\{ \begin{matrix} e^t = u \\ 2\sin\left(t + \tfrac{1}{2}\pi\right) = 2 \\ t^2 - 2 = u^2 - 3 \end{matrix} \right\} \quad \text{so that} \quad t = 0, \quad u = 1.$$

The point of intersection is $(1, 2, -2)$.

$$\mathbf{r}_1'(t) = e^t\mathbf{i} + 2\cos\left(t + \frac{\pi}{2}\right)\mathbf{j} + 2t\mathbf{k}, \quad \mathbf{r}_1'(0) = \mathbf{i}$$

$$\mathbf{r}_2'(u) = \mathbf{i} + 2u\mathbf{k}, \quad \mathbf{r}_2'(1) = \mathbf{i} + 2\mathbf{k}$$

$$\cos\theta = \frac{\mathbf{r}_1'(0)\,\cdot\,\mathbf{r}_2'(1)}{\|\mathbf{r}_1'(0)\|\,\|\mathbf{r}_2'(1)\|} = \frac{1}{5}\sqrt{5} \cong 0.447, \quad \theta \cong 1.11 \text{ radians}$$

19. (a) $\mathbf{r}(t) = a\cos t\,\mathbf{i} + b\sin t\,\mathbf{j}$ (b) $\mathbf{r}(t) = a\cos t\,\mathbf{i} - b\sin t\,\mathbf{j}$

   (c) $\mathbf{r}(t) = a\cos 2t\,\mathbf{i} + b\sin 2t\,\mathbf{j}$ (d) $\mathbf{r}(t) = a\cos 3t\,\mathbf{i} - b\sin 3t\,\mathbf{j}$

21. $\mathbf{r}'(t) = t^3\mathbf{i} + 2t\mathbf{j}$ **23.** $\mathbf{r}'(t) = 2e^{2t}\mathbf{i} - 4e^{-4t}\mathbf{j}$ **25.** $\mathbf{r}'(t) = -2\,\sin t\,\mathbf{i} + 3\,\cos t\,\mathbf{j}$

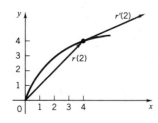   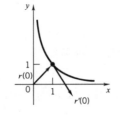   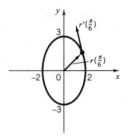

27. $\mathbf{r}(t) = (t^2 + 1)\mathbf{i} + t\mathbf{j}$, $t \geq 1$; or, $\mathbf{r}(t) = \sec^2 t\,\mathbf{i} + \tan t\,\mathbf{j}$, $t \in \left[\tfrac{1}{4}\pi, \tfrac{1}{2}\pi\right)$

29. $\mathbf{r}(t) = \cos t \sin 3t\,\mathbf{i} = \sin t \sin 3t\,\mathbf{j}, \quad t \in [0, \pi]$

31. $y^3 = x^2$

There is no tangent vector at the origin.

33.   We substitute $x = t$, $y = t^2$, $z = t^3$ in the plane equation to obtain

$$4t + 2t^2 + t^3 = 24, \quad (t - 2)\left(t^2 + 4t + 12\right) = 0, \quad t = 2.$$

The twisted cubic intersects the plane at the tip of $\mathbf{r}(2)$, the point $(2, 4, 8)$.

The angle between the curve and the normal line at the point of intersection is the angle between the tangent vector $\mathbf{r}'(2) = \mathbf{i} + 4\mathbf{j} + 12\mathbf{k}$ and the normal $\mathbf{N} = 4\mathbf{i} + 2\mathbf{j} + \mathbf{k}$:

$$\cos\theta = \frac{(\mathbf{i} + 4\mathbf{j} + 12\mathbf{k}) \cdot (4\mathbf{i} + 2\mathbf{j} + \mathbf{k})}{\|\mathbf{i} + 4\mathbf{j} + 12\mathbf{k}\|\,\|4\mathbf{i} + 2\mathbf{j} + \mathbf{k}\|} = \frac{24}{\sqrt{161}\,\sqrt{21}} \cong 0.412, \quad \theta \cong 1.15 \text{ radians.}$$

35. $$\mathbf{r}'(t) = 2\mathbf{j} + 2t\,\mathbf{k}, \quad \|\mathbf{r}'(t)\| = 2\sqrt{1 + t^2}$$

$$\mathbf{T}(t) = \frac{\mathbf{r}'(t)}{\|\mathbf{r}'(t)\|} = \frac{1}{\sqrt{1 + t^2}}\,(\mathbf{j} + t\,\mathbf{k}),$$

$$\mathbf{T}'(t) = \frac{1}{(1 + t^2)^{3/2}}\,[-t\,\mathbf{j} + \mathbf{k}]$$

at $t = 1$:   tip of $\mathbf{r} = (1, 2, 1)$, $\mathbf{T} = \mathbf{T}(1) = \frac{1}{\sqrt{2}}\mathbf{j} + \frac{1}{\sqrt{2}}\mathbf{k}$;

$$\mathbf{T}'(1) = -\frac{1}{2\sqrt{2}}\mathbf{j} + \frac{1}{2\sqrt{2}}\mathbf{k}; \quad \|\mathbf{T}'(1)\| = \frac{1}{2}; \quad \mathbf{N} = \mathbf{N}(1) = \frac{\mathbf{T}'(1)}{\|\mathbf{T}'(1)\|} = -\frac{1}{\sqrt{2}}\mathbf{j} + \frac{1}{2\sqrt{2}}\mathbf{k}$$

normal for osculating plane:

$$\mathbf{T} \times \mathbf{N} = \left(\frac{1}{\sqrt{2}}\mathbf{j} + \frac{1}{\sqrt{2}}\mathbf{k}\right) \times \left(-\frac{1}{\sqrt{2}}\mathbf{j} + \frac{1}{2\sqrt{2}}\mathbf{k}\right) = \frac{1}{2}\mathbf{i}$$

equation for osculating plane:

$$\frac{1}{2}(x - 1) + 0(y - 2) + 0(z - 1) = 0, \quad \text{which gives} \quad x - 1 = 0$$

37.
$$\mathbf{r}'(t) = -2\sin 2t\,\mathbf{i} + 2\cos 2t\,\mathbf{j} + \mathbf{k}, \quad \|\mathbf{r}'(t)\| = \sqrt{5}$$

$$\mathbf{T}(t) = \frac{\mathbf{r}'(t)}{\|\mathbf{r}'(t)\|} = \frac{1}{5}\sqrt{5}\,(-2\sin 2t\,\mathbf{i} + 2\cos 2t\,\mathbf{j} + \mathbf{k})$$

$$\mathbf{T}'(t) = -\tfrac{4}{5}\sqrt{5}\,(\cos 2t\,\mathbf{i} + \sin 2t\,\mathbf{j}), \quad \|\mathbf{T}'(t)\| = \tfrac{4}{5}\sqrt{5}$$

$$\mathbf{N}(t) = \frac{\mathbf{T}'(t)}{\|\mathbf{T}'(t)\|} = -(\cos 2t\,\mathbf{i} + \sin 2t\,\mathbf{j})$$

at $t = \pi/4$:   tip of $\mathbf{r} = (0, 1, \pi/4)$, $\mathbf{T} = \tfrac{1}{5}\sqrt{5}\,(-2\mathbf{i} + \mathbf{k})$, $\mathbf{N} = -\mathbf{j}$

normal for osculating plane:

$$\mathbf{T} \times \mathbf{N} = \tfrac{1}{5}\sqrt{5}\,(-2\mathbf{i} + \mathbf{k}) \times (-\mathbf{j}) = \tfrac{1}{5}\sqrt{5}\,\mathbf{i} + \tfrac{2}{5}\sqrt{5}\,\mathbf{k}$$

equation for osculating plane:

$$\frac{1}{5}\sqrt{5}\,(x - 0) + \frac{2}{5}\sqrt{5}\left(z - \frac{\pi}{4}\right) = 0, \quad \text{which gives} \quad x + 2z = \frac{\pi}{2}$$

39.
$$\mathbf{r}'(t) = \mathbf{i} + 2t\mathbf{j} + 3t^2\mathbf{k}, \quad \|\mathbf{r}'(t)\| = \sqrt{1 + 4t^2 + 9t^4}$$

$$\mathbf{T}(t) = \frac{\mathbf{r}'(t)}{\|\mathbf{r}'(t)\|} = \frac{1}{\sqrt{1 + 4t^2 + 9t^4}}\,(\mathbf{i} + 2t\mathbf{j} + 3t^2\mathbf{k}),$$

$$\mathbf{T}'(t) = \frac{1}{(1 + 4t^2 + 9t^4)^{3/2}}\left[(-4t - 18t^3)\,\mathbf{i} + (2 - 18t^4)\,\mathbf{j} + (6t + 12t^3)\,\mathbf{k}\right]$$

at $t = 1$:   tip of $\mathbf{r} = (1, 1, 1)$, $\mathbf{T} = \dfrac{1}{\sqrt{14}}\,(\mathbf{i} + 2\mathbf{j} + 3\mathbf{k})$,

$$\mathbf{T}' = \frac{1}{7\sqrt{14}}\,(-11\mathbf{i} - 8\mathbf{j} + 9\mathbf{k}), \quad \|\mathbf{T}'\| = \frac{\sqrt{266}}{7\sqrt{14}}, \quad \mathbf{N} = \frac{1}{\sqrt{266}}\,(-11\mathbf{i} - 8\mathbf{j} + 9\mathbf{k})$$

normal for osculating plane:

$$\mathbf{T} \times \mathbf{N} = \frac{1}{\sqrt{14}}\,(\mathbf{i} + 2\mathbf{j} + 3\mathbf{k}) \times \frac{1}{\sqrt{266}}\,(-11\mathbf{i} - 8\mathbf{j} + 9\mathbf{k}) = \frac{\sqrt{19}}{19}\,(3\mathbf{i} - 3\mathbf{j} + \mathbf{k})$$

equation for osculating plane:

$$3(x - 1) - 3(y - 1) + (z - 1) = 0, \quad \text{which gives} \quad 3x - 3y + z = 1$$

41.
$$\mathbf{r}'(t) = e^t\left[(\sin t + \cos t)\,\mathbf{i} + (\cos t - \sin t)\,\mathbf{j} + \mathbf{k}\right], \quad \|\mathbf{r}'(t)\| = e^t\sqrt{3}$$

$$\mathbf{T}(t) = \frac{\mathbf{r}'(t)}{\|\mathbf{r}'(t)\|} = \frac{1}{\sqrt{3}}\left[(\sin t + \cos t)\,\mathbf{i} + (\cos t - \sin t)\,\mathbf{j} + \mathbf{k}\right],$$

$$\mathbf{T}'(t) = \frac{1}{\sqrt{3}}\left[(\cos t - \sin t)\,\mathbf{i} - (\sin t + \cos t)\,\mathbf{j}\right]$$

at $t = 0$:   tip of $\mathbf{r} = (0, 1, 1)$, $\mathbf{T} = \mathbf{T}(0) = \dfrac{1}{\sqrt{3}}(\mathbf{i} + \mathbf{j} + \mathbf{k})$;

$$\mathbf{T}'(0) = \frac{1}{\sqrt{3}}(\mathbf{i} - \mathbf{j}); \quad \|\mathbf{T}'(0)\| = \frac{\sqrt{2}}{\sqrt{3}}; \quad \mathbf{N} = \mathbf{N}(0) = \frac{\mathbf{T}'(0)}{\|\mathbf{T}'(0)\|} = \frac{1}{\sqrt{2}}(\mathbf{i} - \mathbf{j})$$

normal for osculating plane:

$$\mathbf{T} \times \mathbf{N} = \frac{1}{\sqrt{3}}(\mathbf{i} + \mathbf{j} + \mathbf{k}) \times \frac{1}{\sqrt{2}}(\mathbf{i} - \mathbf{j}) = \frac{1}{\sqrt{6}}(\mathbf{i} + \mathbf{j} - 2\mathbf{k})$$

equation for osculating plane:

$$\frac{1}{\sqrt{6}}(x - 0) + \frac{1}{\sqrt{6}}(y - 1) - \frac{2}{\sqrt{6}}(z - 1) = 0, \quad \text{which gives} \quad x + y - 2z + 1 = 0$$

43. $\mathbf{T}_1 = \dfrac{\mathbf{R}'(u)}{\|\mathbf{R}'(u)\|} = -\dfrac{\mathbf{r}'(a + b - u)}{\|\mathbf{r}'(a + b - u)\|} = -\mathbf{T}.$

Therefore $\mathbf{T}'_1(u) = \mathbf{T}'(a + b - u)$ and $\mathbf{N}_1 = \mathbf{N}$.

45.   Let $\mathbf{T}$ be the unit tangent at the tip of $\mathbf{R}(u) = \mathbf{r}(\phi(u))$ as calculated from the parametrization $\mathbf{r}$ and let $\mathbf{T}_1$ be the unit tangent at the same point as calculated from the parametrization $\mathbf{R}$.   Then

$$\mathbf{T}_1 = \frac{\mathbf{R}'(u)}{\|\mathbf{R}'(u)\|} = \frac{\mathbf{r}'(\phi(u))\,\phi'(u)}{\|\mathbf{r}'(\phi(u))\,\phi'(u)\|} = \frac{\mathbf{r}'(\phi(u))}{\|\mathbf{r}'(\phi(u))\|} = \mathbf{T}.$$
$$\phi'(u) > 0$$

This shows the invariance of the unit tangent.

The invariance of the principal normal and the osculating plane follows directly from the invariance of the unit tangent.

47.   (a) Let $t = \Psi(v) = 2\pi - v^2$. When t increases from 0 to 2π, v decreases from $\sqrt{2\pi}$ to 0.

(b) $\mathbf{r}(t) = 2\cos t\,\mathbf{i} + 2\sin t\,\mathbf{j} + 4t\,\mathbf{k}$, $\mathbf{r}'(t) = -2\sin t\,\mathbf{i} + 2\cos t\,\mathbf{j} + 4\,\mathbf{k}$, $\|\mathbf{r}'(t)\| = 2\sqrt{5}$

$$\mathbf{T_r}(t) = -\frac{1}{\sqrt{5}}\sin t\,\mathbf{i} + \frac{1}{\sqrt{5}}\cos t\,\mathbf{j} + \frac{2}{\sqrt{5}}\mathbf{k}, \quad \mathbf{T'_r}(t) = -\frac{1}{\sqrt{5}}\cos t\,\mathbf{i} - \frac{1}{\sqrt{5}}\sin t\,\mathbf{j},$$

$$\|\mathbf{T'_r}(t)\| = 1/\sqrt{5}$$

$$\mathbf{N_r}(t) = -\cos t\,\mathbf{i} - \sin t\,\mathbf{j},$$

$$\mathbf{R}(v) = 2\cos(2\pi - v^2)\,\mathbf{i} + 2\sin(2\pi - v^2)\,\mathbf{j} + 4(2\pi - v^2)\,\mathbf{k}$$

$$\mathbf{R}'(t) = 4v\sin(2\pi - v^2)\,\mathbf{i} - 4v\cos(2\pi - v^2)\,\mathbf{j} + -8v\,\mathbf{k}, \quad \|\mathbf{R}'(t)\| = 4v\sqrt{5}$$

$$\mathbf{T_R}(t) = \frac{1}{\sqrt{5}}\sin(2\pi - v^2)\,\mathbf{i} - \frac{1}{\sqrt{5}}\cos(2\pi - v^2)\,\mathbf{j} - \frac{2}{\sqrt{5}}\mathbf{k}$$

$$\mathbf{T'_R}(t) = -\frac{2v}{\sqrt{5}}\cos(2\pi - v^2)\,\mathbf{i} - \frac{2v}{\sqrt{5}}\sin(2\pi - v^2)\,\mathbf{j}, \quad \|\mathbf{T'_R}(t)\| = 2v/\sqrt{5}$$

$$\mathbf{N_R}(t) = -\cos(2\pi - v^2)\mathbf{i} - \sin(2\pi - v^2)\mathbf{j}$$

$$\mathbf{T_r}(\pi/4) = -\frac{1}{\sqrt{5}}\sin(\pi/4)\mathbf{i} + \frac{1}{\sqrt{5}}\cos(\pi/4)\mathbf{j} + \frac{2}{\sqrt{5}}\mathbf{k} = -\frac{1}{\sqrt{10}}\mathbf{i} + \frac{1}{\sqrt{10}}\mathbf{j} + \frac{2}{\sqrt{5}}\mathbf{k}$$

$$\mathbf{T_R}\left(\sqrt{7}\pi/2\right) = \frac{1}{\sqrt{5}}\sin(2\pi - 7\pi/4)\mathbf{i} - \frac{1}{\sqrt{5}}\cos(2\pi - 7\pi/4)\mathbf{j} - \frac{2}{\sqrt{5}}\mathbf{k} = \frac{1}{\sqrt{10}}\mathbf{i} - \frac{1}{\sqrt{10}}\mathbf{j} - \frac{2}{\sqrt{5}}\mathbf{k} = -\mathbf{T_r}$$

Similarly, $\quad \mathbf{N_r}(\pi/4) = -\frac{1}{\sqrt{2}}\mathbf{i} - \frac{1}{\sqrt{2}}\mathbf{j} = \mathbf{N_R}\left(\sqrt{7}\pi/2\right)$

SECTION 13.4

1. $\mathbf{r}'(t) = \mathbf{i} + t^{1/2}\mathbf{j}, \quad \|\mathbf{r}'(t)\| = \sqrt{1+t}$

$$L = \int_0^8 \sqrt{1+t}\, dt = \left[\frac{2}{3}(1+t)^{3/2}\right]_0^8 = \frac{52}{3}$$

3. $\mathbf{r}'(t) = -a\sin t\,\mathbf{i} + a\cos t\,\mathbf{j} + b\mathbf{k}, \quad \|\mathbf{r}'(t)\| = \sqrt{a^2 + b^2}$

$$L = \int_0^{2\pi} \sqrt{a^2 + b^2}\, dt = 2\pi\sqrt{a^2 + b^2}$$

5. $\mathbf{r}'(t) = \mathbf{i} + \tan t\,\mathbf{j}, \quad \|\mathbf{r}'(t)\| = \sqrt{1 + \tan^2 t} = |\sec t|$

$$L = \int_0^{\pi/4} |\sec t|\, dt = \int_0^{\pi/4} \sec t\, dt = [\ln|\sec t + \tan t|]_0^{\pi/4} = \ln\left(1 + \sqrt{2}\right)$$

7. $\mathbf{r}'(t) = 3t^2\mathbf{i} + 2t\mathbf{j}, \quad \|\mathbf{r}'(t)\| = \sqrt{9t^4 + 4t^2} = |t|\sqrt{4 + 9t^2}$

$$L = \int_0^1 |t\sqrt{4 + 9t2}|\, dt = \int_0^1 t\sqrt{4 + 9t^2}\, dt = \left[\frac{1}{27}(4 + 9t^2)^{3/2}\right]_0^1 = \frac{1}{27}\left(13\sqrt{13} - 8\right)$$

9. $\mathbf{r}'(t) = (\cos t - \sin t)e^t\mathbf{i} + (\sin t + \cos t)e^t\mathbf{j}, \quad \|\mathbf{r}'(t)\| = \sqrt{2}\,e^t$

$$L = \int_0^\pi \sqrt{2}\,e^t\, dt = \sqrt{2}\,(e^\pi - 1)$$

11. $\mathbf{r}'(t) = 2\mathbf{i} + 2t\mathbf{j} - 2t\mathbf{k}, \quad \|\mathbf{r}'(t)\| = 2\sqrt{1 + 2t^2}$

$$L = \int_0^2 2\sqrt{1 + 2t^2}\, dt = \sqrt{2}\int_0^{\tan^{-1}(2\sqrt{2})} \sec^3 u\, du$$

$$(t\sqrt{2} = \tan u)$$

$$= \tfrac{1}{2}\sqrt{2}\,[\sec u\tan u + \ln|\sec u + \tan u|]_0^{\tan^{-1}(2\sqrt{2})} = 6 + \tfrac{1}{2}\sqrt{2}\ln\left(3 + 2\sqrt{2}\right)$$

13. $\mathbf{r}'(t) = \dfrac{1}{t}\mathbf{i} + 2\mathbf{j} + 2t\,\mathbf{k}, \quad \|\mathbf{r}'(t)\| = \sqrt{\dfrac{1}{t^2} + 4 + 4t^2}$

$L = \displaystyle\int_1^e \sqrt{\dfrac{1}{t^2} + 4 + 4t^2}\,dt = \int_1^e \left(\dfrac{1}{t} + 2t\right)dt = \Big[\ln|t| + t^2\Big]_1^e = e^2$

15. $$s = s(t) = \int_a^t \|\mathbf{r}'(u)\|\,du$$

$$s'(t) = \|\mathbf{r}'(t)\| = \|x'(t)\,\mathbf{i} + y'(t)\,\mathbf{j} + z'(t)\,\mathbf{k}\|$$

$$= \sqrt{[x'(t)]^2 + [y'(t)]^2 + [z'(t)]^2}.$$

In the Leibniz notation this translates to

$$\dfrac{ds}{dt} = \sqrt{\left(\dfrac{dx}{dt}\right)^2 + \left(\dfrac{dy}{dt}\right)^2 + \left(\dfrac{dz}{dt}\right)^2}.$$

17. $$s = s(x) = \int_a^x \sqrt{1 + [f'(t)]^2}\,dt$$

$$s'(x) = \sqrt{1 + [f'(x)]^2}.$$

In the Leibniz notation this translates to

$$\dfrac{ds}{dx} = \sqrt{1 + \left(\dfrac{dy}{dx}\right)^2}.$$

19. Let L be the length as computed from $\mathbf{r}$ and L^* the length as computed from $\mathbf{R}$. Then

$$L^* = \int_c^d \|\mathbf{R}'(u)\|\,du = \int_c^d \|\mathbf{r}'(\phi(u))\|\,\phi'(u)\,du = \int_a^b \|\mathbf{r}'(t)\|\,dt = L.$$

$$t = \phi(u)\underline{\quad\quad}$$

21. (a) $s = \displaystyle\int_0^t \sqrt{(-3\sin t)^2 + (3\cos t)^2 + 4^2}\,dt = \int_0^t 5\,dt = 5t$

(b) $t = \dfrac{s}{5}; \qquad R(s) = 3\cos\left(\dfrac{s}{5}\right)\mathbf{i} + 3\sin\left(\dfrac{s}{5}\right)\mathbf{j} + \dfrac{4s}{5}\mathbf{k}$

(c) $\mathbf{r}(t) = 3\cos t\,\mathbf{i} + 3\sin t\,\mathbf{j} + 4t\,\mathbf{k} = 3\mathbf{i} + 0\mathbf{j} + 0\mathbf{k} \implies t = 0$

From part (a), the arc length $s = 5t$ and $5t = 5\pi \implies t = \pi; \quad \mathbf{r}(\pi) = -3\mathbf{i} + 0\,w\mathbf{j} + 4\pi\,\mathbf{k}$

$\implies Q(-3, 0, 4\pi).$

(d) $\mathbf{R}'(s) = -\dfrac{3}{5}\sin\left(\dfrac{s}{5}\right)\mathbf{i} + \dfrac{3}{5}\cos\left(\dfrac{s}{5}\right)\mathbf{j} + \dfrac{4}{5}\mathbf{k}$;

$$\|\mathbf{R}'(s)\| = \sqrt{\left[-\frac{3}{5}\sin\left(\frac{s}{5}\right)\right]^2 + \left[\frac{3}{5}\cos\left(\frac{s}{5}\right)\right]^2 + \left[\frac{4}{5}\right]^2} = 1$$

23. $\mathbf{r}'(t) = t^{3/2}\mathbf{j} + \mathbf{k}$, $\|\mathbf{r}'(t)\| = \sqrt{\left(t^{3/2}\right)^2 + 1} = \sqrt{t^3 + 1}$

$$s = \int_0^{1/2} \sqrt{t^3 + 1}\, dt \cong 0.5077$$

25. $\mathbf{r}'(t) = -3\sin t\,\mathbf{i} + 4\cos t\,\mathbf{j}$, $\|\mathbf{r}'(t)\| = \sqrt{9\sin^2 t + 16\cos^2 t}\, dt$

$$s = \int_0^{2\pi} \sqrt{9\sin^2 t + 16\cos^2 t}\, dt \cong 22.0939$$

SECTION 13.5

1. $\mathbf{r}(t) = r[\cos\theta(t)\,\mathbf{i} + \sin\theta(t)\,\mathbf{j}]$

$\mathbf{r}'(t) = r[-\sin\theta(t)\,\mathbf{i} + \cos\theta(t)\,\mathbf{j}]\theta'(t)$

$\|\mathbf{r}'(t)\| = v \implies r|\theta'(t)| = v \implies |\theta'(t)| = v/r$

$\mathbf{r}''(t) = r[-\cos\theta(t)\,\mathbf{i} - \sin\theta(t)\,\mathbf{j}][\theta'(t)]^2$

$\|\mathbf{r}''(t)\| = r[\theta'(t)]^2 = v^2/r$

3. $\mathbf{r}(t) = at\mathbf{i} + b\sin at\,\mathbf{j}$

$\mathbf{r}'(t) = a\mathbf{i} + ab\cos at\,\mathbf{j}$

$\mathbf{r}''(t) = -a^2 b\sin at\,\mathbf{j}$

$\|\mathbf{r}''(t)\| = a^2|b\sin at|$

$= a^2|y(t)|$

5. $y = \cos\pi x$, $0 \le x \le 2$

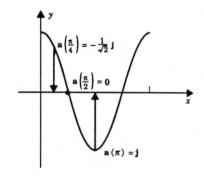

7. $x = \sqrt{1 + y^2}$, $y \ge -1$

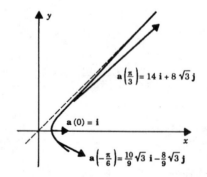

9.   (a)   initial position is tip of $\mathbf{r}(0) = x_0\mathbf{i} + y_0\mathbf{j} + z_0\mathbf{k}$

(b) $\mathbf{r}'(t) = (\alpha\cos\theta)\mathbf{j} + (\alpha\sin\theta - 32t)\mathbf{k}$, $\mathbf{r}'(0) = (\alpha\cos\theta)\mathbf{j} + (\alpha\sin\theta)\mathbf{k}$

(c) $|\mathbf{r}'(0)| = |\alpha|$

(d) $\mathbf{r}''(t) = -32\mathbf{k}$

(e) a parabolic arc from the parabola

$$z = z_0 + (\tan\theta)(y - y_0) - 16\frac{(y - y_0)^2}{\alpha^2 \cos^2\theta}$$

in the plane $x = x_0$

11. (a) $\mathbf{r}'(t) = \dfrac{a\omega}{2}\left(e^{\omega t} - e^{-\omega t}\right)\mathbf{i} + \dfrac{b\omega}{2}\left(e^{\omega t} + e^{-\omega t}\right)\mathbf{j}, \quad \mathbf{r}'(0) = b\omega\mathbf{j}$

(b) $\mathbf{r}''(t) = \dfrac{a\omega^2}{2}\left(e^{\omega t} + e^{-\omega t}\right)\mathbf{i} + \dfrac{b\omega^2}{2}\left(e^{\omega t} - e^{-\omega t}\right)\mathbf{j} = \omega^2\mathbf{r}(t)$

(c) The torque $\boldsymbol{\tau}$ is $\mathbf{0}$: $\boldsymbol{\tau}(t) = \mathbf{r}(t) \times m\mathbf{a}(t) = \mathbf{r}(t) \times m\omega^2\mathbf{r}(t) = \mathbf{0}.$

The angular momentum $\mathbf{L}(t)$ is constant since $\mathbf{L}'(t) = \boldsymbol{\tau}(t) = \mathbf{0}.$

13. We begin with the force equation $\mathbf{F}(t) = \alpha\mathbf{k}$. In general, $\mathbf{F}(t) = m\mathbf{a}(t)$, so that here

$$\mathbf{a}(t) = \frac{\alpha}{m}\mathbf{k}.$$

Integration gives

$$\mathbf{v}(t) = C_1\mathbf{i} + C_2\mathbf{j} + \left(\frac{\alpha}{m}t + C_3\right)\mathbf{k}.$$

Since $\mathbf{v}(0) = 2\mathbf{j}$, we can conclude that $C_1 = 0$, $C_2 = 2$, $C_3 = 0$. Thus

$$\mathbf{v}(t) = 2\mathbf{j} + \frac{\alpha}{m}t\mathbf{k}.$$

Another integration gives

$$\mathbf{r}(t) = D_1\mathbf{i} + (2t + D_2)\mathbf{j} + \left(\frac{\alpha}{2m}t^2 + D_3\right)\mathbf{k}.$$

Since $\mathbf{r}(0) = y_0\mathbf{j} + z_0\mathbf{k}$, we have $D_1 = 0$, $D_2 = y_0$, $D_3 = z_0$, and therefore

$$\mathbf{r}(t) = (2t + y_0)\mathbf{j} + \left(\frac{\alpha}{2m}t^2 + z_0\right)\mathbf{k}.$$

The conditions of the problem require that t be restricted to nonnegative values.
To obtain an equation for the path in Cartesian coordinates, we write out the components

$$x(t) = 0, \quad y(t) = 2t + y_0, \quad z(t) = \frac{\alpha}{2m}t^2 + z_0. \quad (t \geq 0)$$

From the second equation we have

$$t = \tfrac{1}{2}[y(t) - y_0]. \quad (y(t) \geq y_0)$$

Substituting this into the third equation, we get

$$z(t) = \frac{\alpha}{8m}[y(t) - y_0]^2 + z_0. \quad (y(t) \geq y_0)$$

Eliminating t altogether, we have

$$z = \frac{\alpha}{8m}(y - y_0)^2 + z_0. \quad (y \geq y_0)$$

Since $x = 0$, the path of the object is a parabolic arc in the yz-plane.

Answers to (a) through (d):

(a)   velocity: $\mathbf{v}(t) = 2\mathbf{j} + \dfrac{\alpha}{m}\,t\mathbf{k}.$ (b)   speed: $v(t) = \dfrac{1}{m}\sqrt{4m^2 + \alpha^2 t^2}.$

(c)   momentum: $\mathbf{p}(t) = 2m\mathbf{j} + \alpha\,t\mathbf{k}.$

(d)   path in vector form: $\mathbf{r}(t) = (2t + y_0)\mathbf{j} + \left(\dfrac{\alpha}{2m}t^2 + z_0\right)\mathbf{k}, \quad t \geq 0.$

path in Cartesian coordinates: $z = \dfrac{\alpha}{8m}(y - y_0)^2 + z_0, \quad y \geq y_0, \quad x = 0.$

15. $\mathbf{F}(t) = m\,\mathbf{a}(t) = m\,\mathbf{r}''(t) = 2m\mathbf{k}$

17.   From $\mathbf{F}(t) = m\,\mathbf{a}(t)$ we obtain

$$\mathbf{a}(t) = \pi^2[a\cos\pi t\,\mathbf{i} + b\sin\pi t\,\mathbf{j}].$$

By direct calculation using $\mathbf{v}(0) = -\pi b\mathbf{j} + \mathbf{k}$ and $\mathbf{r}(0) = b\mathbf{j}$ we obtain

$$\mathbf{v}(t) = a\pi\sin\pi t\,\mathbf{i} - b\pi\cos\pi t\,\mathbf{j} + \mathbf{k}$$

$$\mathbf{r}(t) = a(1 - \cos\pi t)\mathbf{i} + b(1 - \sin\pi t)\mathbf{j} + t\mathbf{k}.$$

(a) $\mathbf{v}(1) = b\pi\mathbf{j} + \mathbf{k}$ (b) $\|\mathbf{v}(1)\| = \sqrt{\pi^2 b^2 + 1}$

(c) $\mathbf{a}(1) = -\pi^2 a\mathbf{i}$ (d) $m\,\mathbf{v}(1) = m(\pi b\mathbf{j} + \mathbf{k})$

(e) $\mathbf{L}(1) = \mathbf{r}(1) \times m\,\mathbf{v}(1) = [2a\mathbf{i} + b\mathbf{j} + \mathbf{k}] \times [m(b\pi\mathbf{j} + \mathbf{k})]$

$\qquad = m[b(1 - \pi)\mathbf{i} - 2a\mathbf{j} + 2ab\pi\mathbf{k}]$

(f) $\boldsymbol{\tau}(1) = \mathbf{r}(1) \times \mathbf{F}(1) = [2a\mathbf{i} + b\mathbf{j} + \mathbf{k}] \times [-m\pi^2 a\mathbf{i}] = -m\pi^2 a[\mathbf{j} - b\mathbf{k}]$

19.   We have $m\mathbf{v} = m\mathbf{v}_1 + m\mathbf{v}_2$ and $\frac{1}{2}mv^2 = \frac{1}{2}mv_1{}^2 + \frac{1}{2}mv_2{}^2.$

Therefore $\mathbf{v} = \mathbf{v}_1 + \mathbf{v}_2$ and $v^2 = v_1{}^2 + v_2{}^2.$

Since $v^2 = \mathbf{v} \cdot \mathbf{v} = (\mathbf{v}_1 + \mathbf{v}_2) \cdot (\mathbf{v}_1 + \mathbf{v}_2) = v_1{}^2 + v_2{}^2 + 2(\mathbf{v}_1 \cdot \mathbf{v}_2),$

we have $\mathbf{v}_1 \cdot \mathbf{v}_2 = 0$ and $\mathbf{v}_1 \perp \mathbf{v}_2.$

21. $\mathbf{r}''(t) = \mathbf{a}, \quad \mathbf{r}'(t) = \mathbf{v}(0) + t\mathbf{a}, \quad \mathbf{r}(t) = \mathbf{r}(0) + t\,\mathbf{v}(0) + \frac{1}{2}t^2\,\mathbf{a}.$

If neither $\mathbf{v}(0)$ nor $\mathbf{a}$ is zero, the displacement $\mathbf{r}(t) - \mathbf{r}(0)$ is a linear combination of $\mathbf{v}(0)$ and $\mathbf{a}$ and thus remains on the plane determined by these vectors. The equation of this plane can be written

$$[\mathbf{a} \times \mathbf{v}(0)] \cdot [\mathbf{r} - \mathbf{r}(0)] = 0.$$

(If either $\mathbf{v}(0)$ or $\mathbf{a}$ is zero, the motion is restricted to a straight line; if both of these vectors are zero, the particle remains at its initial position $\mathbf{r}(0)$.)

23. $\mathbf{r}(t) = \mathbf{i} + t\mathbf{j} + \left(\dfrac{qE_0}{2m}\right)t^2\mathbf{k}$ **25.** $\mathbf{r}(t) = \left(1 + \dfrac{t^3}{6m}\right)\mathbf{i} + \dfrac{t^4}{12m}\mathbf{j} + t\mathbf{k}$

27.

$$\frac{d}{dt}\left(\frac{1}{2}mv^2\right) = mv\frac{dv}{dt} = m\left(\mathbf{v}\cdot\frac{d\mathbf{v}}{dt}\right) = m\frac{d\mathbf{v}}{dt}\cdot\mathbf{v} = \mathbf{F}\cdot\frac{d\mathbf{r}}{dt}$$

$$= 4r^2\left(\mathbf{r}\cdot\frac{d\mathbf{r}}{dt}\right) = 4r^2\left(r\frac{dr}{dt}\right) = 4r^3\frac{dr}{dt} = \frac{d}{dt}\left(r^4\right).$$

Therefore $d/dt\left(\frac{1}{2}mv^2 - r^4\right) = 0$ and $\frac{1}{2}mv^2 - r^4$ is a constant E. Evaluating E from $t = 0$,   we find that $E = 2m$.

Thus $\frac{1}{2}mv^2 - r^4 = 2m$ and $v = \sqrt{4 + (2/m)\,r^4}$.

SECTION * 13.6

1.   On Earth: year of length T,   average distance from sun d.

On Venus: year of length αT,   average distance from sun $0.72d$.

Therefore

$$\frac{(\alpha T)^2}{T^2} = \frac{(0.72d)^3}{d^3}.$$

This gives $\alpha^2 = (0.72)^3 \cong 0.372$ and $\alpha \cong 0.615$.   Answer: about 61.5% of an Earth year.

3.

$$\left(\frac{dx}{dt}\right)^2 + \left(\frac{dy}{dt}\right)^2 = \left[\frac{d}{dt}(r\cos\theta)\right]^2 + \left[\frac{d}{dt}(r\sin\theta)\right]^2$$

$$= \left[r(-\sin\theta)\frac{d\theta}{dt} + \frac{dr}{dt}\cos\theta\right]^2 + \left[r\cos\theta\frac{d\theta}{dt} + \frac{dr}{dt}\sin\theta\right]^2$$

$$= r^2\sin^2\theta\left(\frac{d\theta}{dt}\right)^2 + \left(\frac{dr}{dt}\right)^2\cos^2\theta + r^2\cos^2\theta\left(\frac{d\theta}{dt}\right)^2 + \left(\frac{dr}{dt}\right)^2\sin^2\theta$$

$$= \left(\frac{dr}{dt}\right)^2 + r^2\left(\frac{d\theta}{dt}\right)^2$$

5.   Substitute

$$r = \frac{a}{1 + e\cos\theta}, \quad \left(\frac{dr}{d\theta}\right)^2 = \left[\frac{-a}{(1 + e\cos\theta)^2}\cdot(-e\sin\theta)\right]^2 = \frac{(ae\sin\theta)^2}{(1 + e\cos\theta)^4}$$

into the right side of the equation and you will see that, with a and e^2 as given, the expression reduces to E.

SECTION 13.7

1. $k = \dfrac{e^{-x}}{\left(1 + e^{-2x}\right)^{3/2}}$

3. $y' = \dfrac{1}{2x^{1/2}}; \quad y'' = \dfrac{-1}{4x^{3/2}}$

$$k = \dfrac{\left|-1/4x^{3/2}\right|}{\left[1 + \left(1/2x^{1/2}\right)^2\right]^{3/2}} = \dfrac{2}{(1 + 4x)^{3/2}}$$

5. $k = \dfrac{\sec^2 x}{\left(1 + \tan^2 x\right)^{3/2}} = |\cos x|$

7. $k = \dfrac{|\sin x|}{\left(1 + \cos^2 x\right)^{3/2}}$

9. $y' = -\dfrac{3}{2} x^{-5/2}; \quad y'' = \dfrac{15}{4} x^{-7/2}$

$$k = \dfrac{\dfrac{15}{4} x^{-7/2}}{\left(1 + \dfrac{9}{4} x^{-5}\right)^{3/2}} = \dfrac{30x^4}{(9 + 4x^5)^{3/2}}$$

11. $k = \dfrac{|x|}{\left(1 + x^4/4\right)^{3/2}}; \quad$ at $\left(2, \dfrac{4}{3}\right), \quad \rho = \dfrac{5}{2}\sqrt{5}$

13. $k = \dfrac{\left|-1/y^3\right|}{\left(1 + 1/y^2\right)^{3/2}} = \dfrac{1}{(1 + y^2)^{3/2}}; \quad$ at $(2, 2), \quad \rho = 5\sqrt{5}$

15. $y'(x) = \dfrac{1}{x+1}, \quad y'(2) = \dfrac{1}{3}; \quad y''(x) = \dfrac{-1}{(x+1)^2}, \quad y''(2) = -\dfrac{1}{9}.$

At $x = 2, \quad k = \dfrac{\left|-\dfrac{1}{9}\right|}{\left[1 + \left(\dfrac{1}{3}\right)^2\right]^{3/2}} = \dfrac{3}{10\sqrt{10}}; \quad \rho = \dfrac{10\sqrt{10}}{3}$

17. $4x^2 + 9y^2 = 36$ is an ellipse which can be parametrized by setting

$$\mathbf{r}(t) = 3\cos t\,\mathbf{i} + 2\sin t\,\mathbf{j}, \quad 0 \le t \le 2\pi$$

The tip of $\mathbf{r}$ is at the point $(3, 0)$ when $t = 0$.

Now

$$x(t) = 3\cos t, \qquad y(t) = 2\sin t$$

and therefore

$$x'(t) = -3\sin t, \quad x''(t) = -3\cos t; \qquad y'(t) = 2\cos t, \quad y''(t) = -2\sin t$$

Also,

$$x'(0) = 0, \quad x''(0) = -3; \qquad y'(0) = 2, \quad y''(0) = 0$$

Thus,

$$k = \frac{|x'(0)y''(0) - y'(0)x''(0)|}{([x'(0)]^2 + [y'(0)]^2)^{3/2}} = \frac{3}{4} \quad \text{and} \quad \rho = \frac{4}{3}$$

19. $k(x) = \dfrac{|-1/x^2|}{(1+1/x^2)^{3/2}} = \dfrac{x}{(x^2+1)^{3/2}}, \quad x > 0$

$k'(x) = \dfrac{(1-2x^2)}{(x^2+1)^{5/2}}, \qquad k'(x) = 0 \implies x = \dfrac{1}{2}\sqrt{2}$

Since k increases on $\left(0, \frac{1}{2}\sqrt{2}\right]$ and decreases on $\left[\frac{1}{2}\sqrt{2}, \infty\right)$, k is maximal at $\left(\frac{1}{2}\sqrt{2}, \frac{1}{2}\ln\frac{1}{2}\right)$.

21. $x(t) = t, \quad x'(t) = 1, \quad x''(t) = 0; \qquad y(t) = \frac{1}{2}t^2, \quad y'(t) = t, \quad y''(t) = 1$

$k = \dfrac{1}{(1+t^2)^{3/2}}$

23. $x(t) = 2t, \quad x'(t) = 2, \quad x''(t) = 0; \qquad y(t) = t^3, \quad y'(t) = 3t^2, \quad y''(t) = 6t$

$k = \dfrac{12|t|}{(4+9t^4)^{3/2}}$

25. $x(t) = e^t\cos t, \quad x'(t) = e^t(\cos t - \sin t), \quad x''(t) = -2e^t\sin t$

$y(t) = e^t\sin t, \quad y'(t) = e^t(\sin t + \cos t), \quad y''(t) = 2e^t\cos t$

$k = \dfrac{|2e^{2t}\cos t\,(\cos t - \sin t) + 2e^{2t}\sin t\,(\cos t + \sin t)|}{[e^{2t}(\cos t - \sin t)^2 + e^{2t}(\cos t + \sin t)^2]^{3/2}} = \dfrac{2e^{2t}}{(2e^{2t})^{3/2}} = \dfrac{1}{2}\sqrt{2}\,e^{-t}$

27. $x(t) = t\cos t, \quad x'(t) = \cos t - t\sin t, \quad x''(t) = -2\sin t - t\cos t$

$y(t) = t\sin t, \quad y'(t) = \sin t + t\cos t, \quad y''(t) = 2\cos t - t\sin t$

$k = \dfrac{|(\cos t - t\sin t)(2\cos t - t\sin t) - (\sin t + t\cos t)(-2\sin t - t\cos t)|}{[(\cos t - t\sin t)^2 + (\sin t + t\cos t)^2]^{3/2}} = \dfrac{2+t^2}{[1+t^2]^{3/2}}$

29. $k = \dfrac{|2/x^3|}{[1+1/x^4]^{3/2}} = \dfrac{2|x^3|}{(x^4+1)^{3/2}}; \quad \text{at } x = \pm 1, \ \rho = \dfrac{2^{3/2}}{2} = \sqrt{2}$

31. We use (13.7.3) and the hint to obtain

$$k = \frac{|ab\sinh^2 t - ab\cosh^2 t|}{[a^2\sinh^2 t + b^2\cosh^2 t]^{3/2}} = \frac{\left|\frac{a}{b}y^2 - \frac{b}{a}x^2\right|}{\left[\left(\frac{ay}{b}\right)^2 + \left(\frac{bx}{a}\right)^2\right]^{3/2}}$$

$$= \frac{a^3b^3\left|\frac{a}{b}y^2 - \frac{b}{a}x^2\right|}{[a^4y^2 + b^4x^2]^{3/2}} = \frac{a^4b^4}{[a^4y^2 + b^4x^2]^{3/2}}.$$

33. By the hint and the fact that $\|\mathbf{T} \times \mathbf{N}\| = 1$,

$$\frac{\|\mathbf{v} \times \mathbf{a}\|}{(ds/dt)^3} = \frac{\left\| \left(\frac{ds}{dt}\mathbf{T}\right) \times \left(\frac{d^2s}{dt^2}\mathbf{T} + k\left(\frac{ds}{dt}\right)^2 \mathbf{N}\right) \right\|}{(ds/dt)^3}$$

$$\mathbf{T} \times \mathbf{T} = 0 \longrightarrow \quad = \frac{\|k(ds/dt)^3(\mathbf{T} \times \mathbf{N})\|}{(ds/dt)^3} = k.$$

35. $\mathbf{r}'(t) = e^t(\cos t - \sin t)\mathbf{i} + e^t(\sin t + \cos t)\mathbf{j} + e^t\mathbf{k}$

$\dfrac{ds}{dt} = \|\mathbf{r}'(t)\| = \sqrt{3}\,e^t, \quad \dfrac{d^2s}{dt^2} = \sqrt{3}\,e^t$

$\mathbf{T}(t) = \dfrac{\mathbf{r}'(t)}{\|\mathbf{r}'(t)\|} = \dfrac{1}{\sqrt{3}}[(\cos t - \sin t)\mathbf{i} + (\sin t + \cos t)\mathbf{j} + \mathbf{k}]$

$\mathbf{T}'(t) = \dfrac{1}{\sqrt{3}}[(-\sin t - \cos t)\mathbf{i} + (\cos t - \sin t)\mathbf{j}]$

Then,
$$k = \frac{\|\mathbf{T}'(t)\|}{ds/dt} = \frac{\sqrt{2/3}}{\sqrt{3}\,e^t} = \frac{1}{3}\sqrt{2}\,e^{-t},$$

$$\mathbf{a_T} = \frac{d^2s}{dt^2} = \sqrt{3}\,e^t, \quad \mathbf{a_N} = k\left(\frac{ds}{dt}\right)^2 = \sqrt{2}\,e^t.$$

37. $\mathbf{r}'(t) = -2\sin 2t\,\mathbf{i} + 2\cos 2t\,\mathbf{j}; \quad \dfrac{ds}{dt} = \|\mathbf{r}'(t)\| = 2, \quad \dfrac{d^2s}{dt^2} = 0$

$\mathbf{T}(t) = \dfrac{\mathbf{r}'(t)}{\|\mathbf{r}'(t)\|} = -\sin 2t\,\mathbf{i} + \cos 2t\,\mathbf{j}$

$\mathbf{T}'(t) = -2(\cos 2t\,\mathbf{i} + \sin 2t\,\mathbf{j})$

Then,
$$k = \frac{\|\mathbf{T}'(t)\|}{ds/dt} = \frac{2}{2} = 1,$$

$$\mathbf{a_T} = \frac{d^2s}{dt^2} = 0, \quad \mathbf{a_N} = k\left(\frac{ds}{dt}\right)^2 = 1 \cdot 4 = 4.$$

39. $\mathbf{r}'(t) = \mathbf{i} + t\mathbf{j} + t^2\mathbf{k}, \quad \dfrac{ds}{dt} = \|\mathbf{r}'(t)\| = \sqrt{t^4 + t^2 + 1}, \quad \dfrac{d^2s}{dt^2} = \dfrac{2t^3 + t}{\sqrt{t^4 + t^2 + 1}}$

$\mathbf{T}(t) = \dfrac{\mathbf{r}'(t)}{\|\mathbf{r}'(t)\|} = \dfrac{1}{\sqrt{t^4 + t^2 + 1}}\left(\mathbf{i} + t\mathbf{j} + t^2\mathbf{k}\right),$

$\mathbf{T}'(t) = \dfrac{1}{(t^4 + t^2 + 1)^{3/2}}\left[-t(2t^2 + 1)\mathbf{i} + (1 - t^4)\mathbf{j} + t(t^2 + 2)\mathbf{k}\right].$

Then,
$$k = \frac{\|\mathbf{T}'(t)\|}{ds/dt} = \frac{\sqrt{t^2(2t^2 + 1)^2 + (1 + t^4)^2 + t^2(t^2 + 2)^2}}{(t^4 + t^2 + 1)^2}$$

$$= \frac{\sqrt{(t^4 + 4t^2 + 1)(t^4 + t^2 + 1)}}{(t^4 + t^2 + 1)^2} = \frac{\sqrt{t^4 + 4t^2 + 1}}{(t^4 + t^2 + 1)^{3/2}},$$

$$\mathbf{a_T} = \frac{d^2s}{dt^2} = \frac{2t^3 + t}{\sqrt{t^4 + t^2 + 1}}, \quad \mathbf{a_N} = k\left(\frac{ds}{dt}\right)^2 = \frac{\sqrt{t^4 + 4t^2 + 1}}{\sqrt{t^4 + t^2 + 1}}.$$

41. By Exercise 40

$$k = \frac{\left|\left(e^{a\theta}\right)^2 + 2\left(ae^{a\theta}\right)^2 - \left(e^{a\theta}\right)\left(a^2 e^{a\theta}\right)\right|}{\left[\left(e^{a\theta}\right)^2 + \left(ae^{a\theta}\right)^2\right]^{3/2}} = \frac{e^{-a\theta}}{\sqrt{1+a^2}}.$$

43. By Exercise 40

$$k = \frac{\left|a^2(1-\cos\theta)^2 + 2a^2\sin^2\theta - a^2(1-\cos\theta)(\cos\theta)\right|}{\left[a^2(1-\cos\theta)^2 + a^2\sin^2\theta\right]^{3/2}}$$

$$= \frac{3a^2(1-\cos\theta)}{[2a^2(1-\cos\theta)]^{3/2}} = \frac{3ar}{[2ar]^{3/2}} = \frac{3}{2\sqrt{2ar}}.$$

45. (a) For $0 \le \theta \le \pi$,

$$s(\theta) = \int_\theta^\pi \sqrt{[x'(t)]^2 + [y'(t)]^2}\, dt = \int_\theta^\pi \sqrt{R^2(1-\cos t)^2 + R^2\sin^2 t}\, dt$$

$$= \int_\theta^\pi R\sqrt{2(1-\cos t)}\, dt = \int_\theta^\pi 2R\sin\frac{1}{2}t\, dt = 4R\cos\frac{1}{2}\theta = 4R\left|\cos\frac{1}{2}\theta\right|.$$

For $\pi \le \theta \le 2\pi$,

$$s(\theta) = \int_\pi^\theta 2R\sin\frac{1}{2}t\, dt = -4R\cos\frac{1}{2}\theta = 4R\left|\cos\frac{1}{2}\theta\right|.$$

(b) $$k(\theta) = \frac{|x'(\theta)y''(\theta) - y'(\theta)x''(\theta)|}{\{[x'(\theta)]^2 + [y'(\theta)]\}^{3/2}} = \frac{|R(1-\cos\theta)R\cos\theta - R\sin\theta\,(R\sin\theta)|}{8R^3\sin^3\frac{1}{2}\theta}.$$

This reduces to $k(\theta) = 1/(4R\sin\frac{1}{2}\theta)$ and gives $\rho(\theta) = 4R\sin\frac{1}{2}\theta.$

(c) $\rho^2 + s^2 = 16R^2$

47. Straightforward calculation gives

$$s(\theta) = 4a\left|\cos\frac{1}{2}\theta\right| \quad \text{and} \quad \rho(\theta) = \frac{4}{3}a\sin\frac{1}{2}\theta.$$

Therefore

$$9\rho^2 + s^2 = 16a^2.$$

PROJECTS AND EXPLORATIONS

13.1. $\mathbf{f}(t) = t^3\,\mathbf{i} - \sin t\,\mathbf{j} + \ln t\,\mathbf{k};$ $\mathbf{f}'(t)(1) \cong 3\,\mathbf{i} - 0.540302\,\mathbf{j} + 1\,\mathbf{k}$

t	t^3	$-\sin t$	$\ln t$
1	1	−0.841471	0
1.1	1.331	−0.891207	0.0953102
1.01	1.030301	−0.846832	0.00099503
1.001	1.003003	−0.842011	0.0009995
1.0001	1.000300	−0.841525	0.0001000
0.9999	0.999700	−0.841417	−0.000100
0.999	0.997003	−0.840930	−0.001001
0.99	0.990299	−0.836026	−0.010050
0.9	0.729	−0.783327	−0.105361

h		Diff. Quot		Euc. Norm	Sup Norm
0.1	3.31	−0.497634	0.9531018	0.316454	0.31
0.01	3.0301	−0.536086	0.9950331	0.030797	0.0301
0.001	3.003001	−0.539881	0.9995003	0.003071	0.003001
0.0001	3.000300	−0.540260	0.9999500	0.000307	0.000300

h		Symm. Diff.		Euc. Norm	Sup Norm
0.1	3.01	−0.539402	1.003353	0.0105856	0.01
0.01	3.0001	−0.540293	1.000033	0.0001058	0.0010000
0.001	3.000001	−0.540302	1.000000	0.0000011	0.0000010
0.0001	3.000000	−0.540302	1.000000	0.000000	0.000000

$\mathbf{f}(t) = \sec t\,\mathbf{i} + e^{3t}\,\mathbf{j} - (3 + \tan^3 t)^6\,\mathbf{k};$ $\mathbf{f}'(t)(0.75) \cong 1.273214,\mathbf{i} + 28.46321\,\mathbf{j} - 5935.44\,\mathbf{k}$

t	$\sec t$	e^{3t}	$-(3 + \tan^3 t)^6$
0.75	1.366701	9.487736	−1682.77
0.85	1.515190	12.80710	−2586.922
0.76	1.379622	9.776680	−1744.31
0.751	1.367976	9.516242	−1688.73
0.7501	1.366828	9.490583	−1683.37
0.7499	0.366574	9.484980	−1682.18
0.749	0.365430	9.549315	−1676.86
0.74	0.354154	9.207331	−1625.47
0.65	0.256149	7.028688	−1249.19

h		Diff. Quot		Euc. Norm	Sup Norm
0.1	1.484891	33.19368	−9041.47	3106.041	3106.037
0.01	1.292128	28.89446	−6153.77	218.3339	218.3335
0.001	1.275086	28.50595	−5956.64	21.20052	21.20048
0.0001	1.273401	28.46748	−5937.55	2.113924	2.113920

h		Symm. Diff.		Euc. Norm	Sup Norm
0.1	1.295205	28.89208	−6688.66	753.2247	753.2246
0.01	1.273431	28.46748	−5942.23	6.792408	6.972047
0.001	1.273216	28.46325	−5935.51	0.0678569	0.0678569
0.0001	1.273214	28.46321	−5935.44	0.0006786	0.0006786

Convergence in the Euclidean and sup norms is as expected — linear for the difference quotient and quadratic for the symmetric difference. The sup norm may be easier to compute than the Euclidean norm.

13.3.   (a)

n	arc length
1	1.6456220175
2	1.8421339398
4	1.8944267528
8	1.9056285458
16	1.9084335872
32	1.9091352106

(b) arc length $\cong 1.9093692063;$ err $\cong (0.2631376728)\, n^{-2.0340432057}$

(c)

n	Trap.	Simpson
1	3.2708894641	2.0014148248
2	2.3187834846	1.9069501674
4	2.0099084968	1.0905241472
8	1.9346127346	1.9093779048
16	1.9156866123	1.9093697510
32	1.9109489663	1.9093692403

The order of accuracy seems to be: Simpson, summation, trapezoid.

(d) The summations certainly seem to be increasing with the number of points.

(e) Another parametrization is: $\mathbf{R}(u) = \left(\sin^{-1}[-u]\right)^5 \mathbf{i} + \sin^{-1}(-u)\mathbf{j} + u\,\mathbf{k}, \quad u \in [-\sin(1), 0]$

CHAPTER 14

SECTION 14.1

1.   dom (f) = the first and third quadrants, including the axes;   ran $(f) = [0, \infty)$

3.   dom (f) = the set of all points (x, y) except those on the line $y = -x$;   ran $(f) = (-\infty, 0) \cup (0, \infty)$

5.   dom (f) = the entire plane;   ran $(f) = (-1, 1)$ since

$$\frac{e^x - e^y}{e^x + e^y} = \frac{e^x + e^y - 2e^y}{e^x + e^y} = 1 - \frac{2}{e^{x-y} + 1}$$

and the last quotient takes on all values between 0 and 2.

7.   dom (f) = the first and third quadrants, excluding the axes;   ran $(f) = (-\infty, \infty)$

9.   dom (f) = the set of all points (x, y) with $x^2 < y$ —in other words, the set of all points of the plane above the parabola $y = x^2$;   ran $(f) = (0, \infty)$

11. dom (f) = the set of all points (x, y) with $-3 \le x \le 3$, $-2 \le y \le 2$ (a rectangle);   ran $(f) = [-2, 3]$

13. dom (f) = the set of all points (x, y, z) not on the plane $x + y + z = 0$;   ran $(f) = \{-1, 1\}$

15. dom (f) = the set of all points (x, y, z) with $|y| < |x|$;   ran $(f) = (-\infty, 0]$

17. dom (f) = the set of all points (x, y) with $x^2 + y^2 < 9$ —in other words, the set of all points of the plane inside the circle $x^2 + y^2 = 9$;   ran $(f) = [2/3, \infty)$

19. dom (f) = the set of all points (x, y, z) with $x + 2y + 3z > 0$ —in other words, the set of all points in space that lie on the same side of the plane $x + 2y + 3z = 0$ as the point $(1, 1, 1)$;   ran $(f) = (-\infty, \infty)$

21. dom (f) = all of space;   ran $(f) = (0, \infty)$

23. dom $(f) = \{x : x \ge 0\}$;   range $(f) = [0, \infty)$

dom $(g) = \{(x, y) : x \ge 0, \ y \text{ real}\}$;   range $(g) = [0, \infty)$

dom $(h) = \{(x, y, z) : x \ge 0, \ y, z \text{ real}\}$;   range $(h) = [0, \infty)$

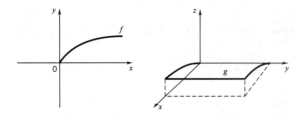

25. $\displaystyle\lim_{h \to 0} \frac{f(x+h, y) - f(x, y)}{h} = \lim_{h \to 0} \frac{2(x+h)^2 - y - (2x^2 - y)}{h} = \lim_{h \to 0} \frac{4xh + 2h^2}{h} = 4x$

$\displaystyle\lim_{h \to 0} \frac{f(x, y+h) - f(x, y)}{h} = \lim_{h \to 0} \frac{2x^2 - (y+h) - (2x^2 - y)}{h} = -1$

27. $\displaystyle\lim_{h \to 0} \frac{f(x+h, y) - f(x, y)}{h} = \lim_{h \to 0} \frac{3(x+h) - (x+h)y + 2y^2 - (3x - xy + 2y^2)}{h} = \lim_{h \to 0} \frac{3h - hy}{h} = 3 - y$

$$\lim_{h\to 0}\frac{f(x,y+h)-f(x,y)}{h}=\lim_{h\to 0}\frac{3x-x(y+h)+2(y+h)^2-(3x-xy+2y^2)}{h}$$
$$=\lim_{h\to 0}\frac{-xh+4yh+2h^2}{h}=-x+4y$$

29.
$$\lim_{h\to 0}\frac{f(x+h,y)-f(x,y)}{h}=\lim_{h\to 0}\frac{\cos[(x+h)y]-\cos[xy]}{h}$$
$$=\lim_{h\to 0}\frac{\cos[xy]\cos[hy]-\sin[xy]\sin[hy]-\cos[xy]}{h}$$
$$=\cos[xy]\left(\lim_{h\to 0}\frac{\cos[hy]-1}{h}\right)-\sin[xy]\lim_{h\to 0}\frac{\sin hy}{h}$$
$$=y\cos[xy]\left(\lim_{h\to 0}\frac{\cos[hy]-1}{hy}\right)-y\sin[xy]\lim_{h\to 0}\frac{\sin hy}{hy}$$
$$=-y\sin[xy]$$

$$\lim_{h\to 0}\frac{f(x,y+h)-f(x,y)}{h}=\lim_{h\to 0}\frac{\cos[x(y+h)]-\cos[xy]}{h}$$
$$=\lim_{h\to 0}\frac{\cos[xy]\cos[hx]-\sin[xy]\sin[hx]-\cos[xy]}{h}$$
$$=\cos[xy]\left(\lim_{h\to 0}\frac{\cos[hx]-1}{h}\right)-\sin[xy]\lim_{h\to 0}\frac{\sin hx}{h}$$
$$=x\cos[xy]\left(\lim_{h\to 0}\frac{\cos[hx]-1}{hx}\right)-x\sin[xy]\lim_{h\to 0}\frac{\sin hx}{hx}$$
$$=-x\sin[xy]$$

31.   (a) $f(x,y)=x^2y$ (b) $f(x,y)=\pi x^2y$ (c) $f(x,y)=|2\mathbf{i}\times(x\mathbf{i}+y\mathbf{j})|=2|y|$

33.   Surface area: $S=2lw+2lh+2hw=20 \implies w=\dfrac{20-2lh}{2l+2h}=\dfrac{10-lh}{l+h}$

Volume: $V=lwh=\dfrac{lh(10-lh)}{l+h}$

35. $V=\pi r^2h+\dfrac{4}{3}\pi r^3$

SECTION 14.2

1.   a quadric cone

3.   a parabolic cylinder

5.   a hyperboloid of one sheet

7.   a sphere

9.   an elliptic paraboloid

11.   a hyperbolic paraboloid

13.
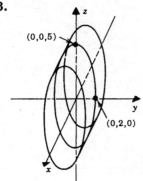

15.
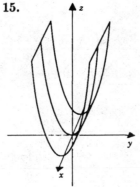

17.
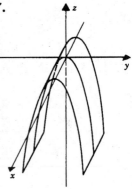

19.
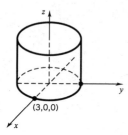

21.
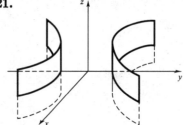

23.
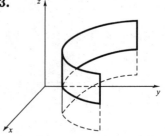

25. elliptic paraboloid
xy-trace: the origin
xz-trace: the parabola $x^2 = 4z$
yz-trace: the parabola $y^2 = 9z$
surface has the form of Figure 14.2.5

27. quadric cone
xy-trace: the origin
xz-trace: the lines $x = \pm 2z$
yz-trace: the lines $y = \pm 3z$
surface has the form of Figure 14.2.4

29. hyperboloid of two sheets
xy-trace: none
xz-trace: the hyperbola $4z^2 - x^2 = 4$
yz-trace: the hyperbola $9z^2 - y^2 = 9$
surface has the form of Figure 14.2.3

31. hyperboloid of two sheets
xy-trace: the hyperbola $4x^2 - 9y^2 = 36$
xz-trace: the hyperbola $x^2 - 4z^2 = 4$
yz-trace: none
see Figure 14.2.3

33. elliptic paraboloid
xy-trace: the origin
xz-trace: the parabola $x^2 = 4z$
yz-trace: the parabola $y^2 = 9z$
surface has the form of Figure 14.2.5

35. hyperboloid of two sheets
xy-trace: the hyperbola $9y^2 - 4x^2 = 36$
xz-trace: none
yz-trace: the hyperbola $y^2 - 4z^2 = 4y$
see Figure 14.2.3

37. paraboloid of revolution
xy-trace: the origin
xz-trace: the parabola $x^2 = 4z$
yz-trace: the parabola $y^2 = 4z$
surface has the form of Figure 14.2.5

39. (a) an elliptic paraboloid (vertex down if A and B are both positive, vertex up if A and B are both negative)

(b) a hyperbolic paraboloid

(c) the xy-plane if A and B are both zero; otherwise a parabolic cylinder

41. $x^2 + y^2 - 4z = 0$ (paraboloid of revolution)

43. (a) a circle

(b) (i) $\sqrt{x^2 + y^2} = -3z$ (ii) $\sqrt{x^2 + z^2} = \frac{1}{3}y$

45. $x + 2y + 3\left(\dfrac{x + y - 6}{2}\right) = 6$ or $5x + 7y = 30$, a line

47. $\left.\begin{array}{l} x^2 + y^2 + (z-1)^2 = \frac{3}{2} \\ x^2 + y^2 - z^2 = 1 \end{array}\right\}$ $(z^2 + 1) + (z - 1)^2 = \frac{3}{2}$; $(2z - 1)^2 = 0$, $z = \frac{1}{2}$ so that $x^2 + y^2 = \frac{5}{4}$

49. $x^2 + y^2 + (x^2 + 3y^2) = 4$ or $x^2 + 2y^2 = 2$, an ellipse

51. $x^2 + y^2 = (2 - y)^2$ or $x^2 = -4(y - 1)$, a parabola

SECTION 14.3

1. lines of slope 1: $y = x - c$

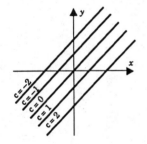

3. parabolas: $y = x^2 - c$

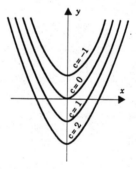

5. the y-axis and the lines $y = \left(\dfrac{1 - c}{c}\right)x$

with the origin omitted throughout

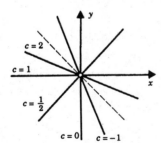

7. the cubics $y = x^3 - c$

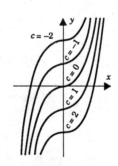

9. the lines $y = \pm x$ and the hyperbolas $x^2 - y^2 = c$

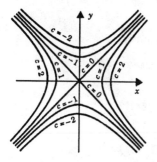

11. pairs of horizontal lines $y = \pm\sqrt{c}$ and the x-axis

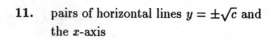

13. the circles $x^2 + y^2 = e^c$, c real

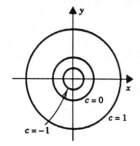

15. the curves $y = e^{cx^2}$ with the point $(0,1)$ omitted

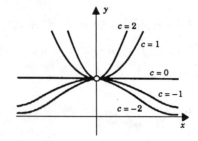

17. the coordinate axes and pairs of lines

$$y = \pm\frac{\sqrt{1-c}}{\sqrt{c}}\, x, \text{ the origin}$$

omitted throughout

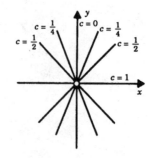

19. $x + 2y + 3z = 0$,   plane through the origin

21. $z = \sqrt{x^2 + y^2}$, the upper nappe of the circular cone $z^2 = x^2 + y^2$ (Figure 14.2.4)

23. the elliptic paraboloid $\dfrac{x^2}{(1/2)^2} + \dfrac{y^2}{(1/3)^2} = 72z$ (Figure 14.2.5)

25.   (i)   hyperboloid of two sheets   (Figure 14.2.3)

(ii)   circular cone   (Figure 14.2.4)

(iii)   hyperboloid of one sheet   (Figure 14.2.2)

27. The level curves of f are: $1 - 4x^2 - y^2 = c$. Substituting $P(0,1)$ into this equation,

we have

$$1 - 4(0)^2 - (1)^2 = c \implies c = 0$$

The level curve that contains P is: $1 - 4x^2 - y^2 = 0$, or $4x^2 + y^2 = 1$.

29. The level curves of f are: $y^2 \tan^{-1} x = c$. Substituting $P(1,2)$ into this equation,

we have

$$4^2 \tan^{-1} 1 = c \implies c = \pi$$

The level curve that contains P is: $y^2 \tan^{-1} x = \pi$.

31. The level surfaces of f are: $x^2 + 2y^2 - 2xyz = c$. Substituting $P(-1,2,1)$ into this equation,

we have

$$(-1)^2 + 2(2)^2 - 2(-1)(2)(1) = c \implies c = 13$$

The level surface that contains P is: $x^2 + 2y^2 - 2xyz = 13$.

33. $\dfrac{GmM}{x^2 + y^2 + z^2} = c \implies x^2 + y^2 + z^2 = \dfrac{GmM}{c}$; the surfaces of constant gravitational force
are concentric spheres.

35. (a) $T(x,y) = \dfrac{k}{x^2 + y^2}$, where k is a constant.

(b) $\dfrac{k}{x^2 + y^2} = c \implies x^2 + y^2 = \dfrac{k}{c}$; the level curves are concentric circles.

(c) $T(1,2) = \dfrac{k}{1^2 + 2^2} = 50 \implies k = 250 \implies T(x,y) = \dfrac{250}{x^2 + y^2}$

Now, $T(3,4) = \dfrac{250}{3^2 + 4^2} = 10°$

37. $f(x,y) = y^2 - y^3$; F   **39.** $f(x,y) = \cos\sqrt{x^2 + y^2}$; A   **41.** $f(x,y) = xye^{-(x^2+y^2)/2}$; E

SECTION 14.4

1. $\dfrac{\partial f}{\partial x} = 6x - y$, $\dfrac{\partial f}{\partial y} = 1 - x$ **3.** $\dfrac{\partial \rho}{\partial \phi} = \cos\phi\cos\theta$, $\dfrac{\partial \rho}{\partial \theta} = -\sin\phi\sin\theta$

5. $\dfrac{\partial f}{\partial x} = e^{x-y} + e^{y-x}$, $\dfrac{\partial f}{\partial y} = -e^{x-y} - e^{y-x}$ **7.** $\dfrac{\partial g}{\partial x} = \dfrac{(AD - BC)y}{(Cx + Dy)^2}$, $\dfrac{\partial g}{\partial y} = \dfrac{(BC - AD)x}{(Cx + Dy)^2}$

9. $\dfrac{\partial u}{\partial x} = y + z$, $\dfrac{\partial u}{\partial y} = x + z$, $\dfrac{\partial u}{\partial z} = x + y$

11. $\dfrac{\partial f}{\partial x} = z\cos(x - y)$, $\dfrac{\partial f}{\partial y} = -z\cos(x - y)$, $\dfrac{\partial f}{\partial z} = \sin(x - y)$

13. $\dfrac{\partial \rho}{\partial \theta} = e^{\theta+\phi}\left[\cos(\theta-\phi) - \sin(\theta-\phi)\right], \quad \dfrac{\partial \rho}{\partial \phi} = e^{\theta+\phi}\left[\cos(\theta-\phi) + \sin(\theta-\phi)\right]$

15. $\dfrac{\partial f}{\partial x} = 2xy \sec xy + x^2 y(\sec xy)(\tan xy)y = 2xy \sec xy + x^2 y^2 \sec xy \tan xy$

$\dfrac{\partial f}{\partial y} = x^2 \sec xy + x^2 y(\sec xy)(\tan xy)x = x^2 \sec xy + x^3 y \sec xy \tan xy$

17. $\dfrac{\partial h}{\partial x} = \dfrac{x^2 + y^2 - x(2x)}{(x^2+y^2)^2} = \dfrac{y^2 - x^2}{(x^2+y^2)^2}$

$\dfrac{\partial h}{\partial y} = \dfrac{-2xy}{(x^2+y^2)^2}$

19. $\dfrac{\partial f}{\partial x} = \dfrac{(y \cos x)\sin y - (x \sin y)(-y \sin x)}{(y \cos x)^2} = \dfrac{\sin y(\cos x + x \sin x)}{y \cos^2 x}$

$\dfrac{\partial f}{\partial y} = \dfrac{(y \cos x)(x \cos y) - (x \sin y)\cos x}{(y \cos x)^2} = \dfrac{x(y \cos y - \sin y)}{y^2 \cos x}$

21. $\dfrac{\partial h}{\partial x} = 2f(x)f'(x)g(y), \quad \dfrac{\partial h}{\partial y} = [f(x)]^2 g'(y)$

23. $\dfrac{\partial f}{\partial x} = (y^2 \ln z)z^{xy^2}, \quad \dfrac{\partial f}{\partial y} = (2xy \ln z)z^{xy^2}, \quad \dfrac{\partial f}{\partial z} = xy^2 z^{xy^2-1}$

25. $\dfrac{\partial h}{\partial r} = 2re^{2t}\cos(\theta - t) \qquad\qquad \dfrac{\partial h}{\partial \theta} = -r^2 e^{2t}\sin(\theta - t)$

$\dfrac{\partial h}{\partial t} = 2r^2 e^{2t}\cos(\theta - t) + r^2 e^{2t}\sin(\theta - t) = r^2 e^{2t}[2\cos(\theta - t) + \sin(\theta - t)]$

27. $\dfrac{\partial f}{\partial x} = z\,\dfrac{1}{1+(y/x)^2}\left(\dfrac{-y}{x^2}\right) = -\dfrac{yz}{x^2+y^2}$

$\dfrac{\partial f}{\partial y} = z\,\dfrac{1}{1+(y/x)^2}\left(\dfrac{1}{x}\right) = \dfrac{xz}{x^2+y^2}$

$\dfrac{\partial f}{\partial x} = \tan^{-1}(y/x)$

29. $f_x(x,y) = e^x \ln y, \quad f_x(0,e) = 1; \quad f_y(x,y) = \dfrac{1}{y}e^x, \quad f_y(0,e) = e^{-1}$

31. $f_x(x,y) = \dfrac{y}{(x+y)^2}, \quad f_x(1,2) = \dfrac{2}{9}; \quad f_y(x,y) = \dfrac{-x}{(x+y)^2}, \quad f_y(1,2) = -\dfrac{1}{9}$

33. $f_x(x,y) = \lim\limits_{h\to 0}\dfrac{(x+h)^2 y - x^2 y}{h} = \lim\limits_{h\to 0} y\left(\dfrac{2xh+h^2}{h}\right) = y\lim\limits_{h\to 0}(2x+h) = 2xy$

$f_x(x,y) = \lim\limits_{h\to 0}\dfrac{x^2(y+h) - x^2 y}{h} = \lim\limits_{h\to 0}\dfrac{x^2 h}{h} = \lim\limits_{h\to 0} x^2 = x^2$

35.
$$f_x(x,y) = \lim_{h\to 0} \frac{\ln\left(y(x+h)^2\right) - \ln x^2 y}{h} = \lim_{h\to 0} \frac{\ln y + 2\ln(x+h) - 2\ln x - \ln y}{h}$$

$$= 2\lim_{h\to 0} \frac{\ln(x+h) - \ln x}{h} = 2\frac{d}{dx}(\ln x) = \frac{2}{x}$$

$$f_y(x,y) = \lim_{h\to 0} \frac{\ln\left(x^2(y+h)\right) - \ln x^2 y}{h} = \lim_{h\to 0} \frac{\ln x^2 + \ln(y+h) - \ln x^2 - \ln y}{h}$$

$$= \lim_{h\to 0} \frac{\ln(y+h) - \ln y}{h} = \frac{d}{dy}(\ln y) = \frac{1}{y}$$

37.
$$f_x(x,y) = \lim_{h\to 0} \frac{1}{h}\left\{\frac{1}{(x+h)-y} - \frac{1}{x-y}\right\} = \lim_{h\to 0} \frac{1}{h}\left\{\frac{-h}{(x+h-y)(x-y)}\right\}$$

$$= \lim_{h\to 0} \frac{-1}{(x+h-y)(x-y)} = \frac{-1}{(x-y)^2}$$

$$f_x(x,y) = \lim_{h\to 0} \frac{1}{h}\left\{\frac{1}{x-(y+h)} - \frac{1}{x-y}\right\} = \lim_{h\to 0} \frac{1}{h}\left\{\frac{h}{(x-y-h)(x-y)}\right\}$$

$$= \lim_{h\to 0} \frac{1}{(x-y-h)(x-y)} = \frac{1}{(x-y)^2}$$

39.
$$f_x(x,y,z) = \lim_{h\to 0} \frac{(x+h)y^2 z - xy^2 z}{h} = \lim_{h\to 0} y^2 z = y^2 z$$

$$f_y(x,y,z) = \lim_{h\to 0} \frac{x(y+h)^2 z - xy^2 z}{h} = \lim_{h\to 0} \frac{xz(2yh+h^2)}{h}$$

$$= \lim_{h\to 0} xz(2y+h) = 2xyz$$

$$f_z(x,y,z) = \lim_{h\to 0} \frac{xy^2(x+h) - xy^2 z}{h} = \lim_{h\to 0} xy^2 = xy^2$$

41. (b) The slope of the tangent line to C at the point $P(x_0, y_0, f(x_0, y_0))$ is $f_y(x_0, y_0)$

Thus, equations for the tangent line are:

$$y = y_0, \quad z - z_0 = f_y(x_0, y_0)(y - y_0)$$

43. Let $z = f(x,y) = x^2 + y^2$. Then $f(2,1) = 5$, $f_y(x,y) = 2y$ and $f_y(2,1) = 2$

By Exercise 41, equations for the tangent line are:

$$x = 2, \quad z - 5 = 2(y-1)$$

45. Let $z = f(x,y) = \dfrac{x^2}{y^2 - 3}$. Then $f(3,2) = 9$, $f_x(x,y) = \dfrac{2x}{y^2 - 3}$ and $f_x(3,2) = 6$

By Exercise 41, equations for the tangent line are:

$$y = 2, \quad z - 9 = 6(x - 3)$$

47. $u_x(x, y) = 2x = v_y(x, y); \qquad u_y(x, y) = -2y = -v_x(x, y)$

49. $u_x(x, y) = \dfrac{1}{2} \dfrac{1}{x^2 + y^2} 2x = \dfrac{x}{x^2 + y^2}; \quad v_y(x, y) = \dfrac{1}{1 + (y/x)^2}\left(\dfrac{1}{x}\right) = \dfrac{x}{x^2 + y^2}$

Thus, $u_x(x, y) = v_y(x, y)$.

$u_y(x, y) = \dfrac{1}{2} \dfrac{1}{x^2 + y^2} 2y = \dfrac{y}{x^2 + y^2}; \quad v_x(x, y) = \dfrac{1}{1 + (y/x)^2}\left(\dfrac{-y}{x^2}\right) = \dfrac{-y}{x^2 + y^2}$

Thus, $u_y(x, y) = -v_x(x, y)$.

51.   (a) f depends only on y. \qquad\qquad (b) f depends only on x.

53.   (a) $50\sqrt{3}$ in.^{2}

(b) $\dfrac{\partial A}{\partial b} = \dfrac{1}{2} c \sin\theta;$ at time t_0, $\dfrac{\partial A}{\partial b} = 5\sqrt{3}$

(c) $\dfrac{\partial A}{\partial \theta} = \dfrac{1}{2} bc \cos\theta;$ at time t_0, $\dfrac{\partial A}{\partial \theta} = 50$

(d) with $h = \dfrac{\pi}{180}, \quad A(b, c, \theta + h) - A(b, c, \theta) \cong h\dfrac{\partial A}{\partial \theta} = \dfrac{\pi}{180}(50) = \dfrac{5\pi}{18}$ in.^{2}

(e) $0 = \dfrac{1}{2} \sin\theta \left(b\dfrac{\partial c}{\partial b} + c\right);$ at time t_0, $\dfrac{\partial c}{\partial b} = \dfrac{-c}{b} = -2$

55.   (a) y_0-section: $\mathbf{r}(x) = x\mathbf{i} + y_0\mathbf{j} + f(x, y_0)\mathbf{k}$

tangent line: $\mathbf{R}(t) = [x_0\mathbf{i} + y_0\mathbf{j} + f(x_0, y_0)\mathbf{k}] + t\left[\mathbf{i} + \dfrac{\partial f}{\partial x}(x_0, y_0)\mathbf{k}\right]$

(b) x_0-section: $\mathbf{r}(y) = x_0\mathbf{i} + y\mathbf{j} + f(x_0, y)\mathbf{k}$

tangent line: $\mathbf{R}(t) = [x_0\mathbf{i} + y_0\mathbf{j} + f(x_0, y_0)\mathbf{k}] + t\left[\mathbf{j} + \dfrac{\partial f}{\partial y}(x_0, y_0)\mathbf{k}\right]$

(c) For (x, y, z) in the plane

$[(x - x_0)\mathbf{i} + (y - y_0)\mathbf{j} + (z - f(x_0, y_0))\mathbf{k}] \cdot \left[\left(\mathbf{i} + \dfrac{\partial f}{\partial x}(x_0, y_0)\mathbf{k}\right) \times \left(\mathbf{j} + \dfrac{\partial f}{\partial y}(x_0, y_0)\mathbf{k}\right)\right] = 0.$

From this it follows that

$$z - f(x_0, y_0) = (x - x_0)\dfrac{\partial f}{\partial x}(x_0, y_0) + (y - y_0)\dfrac{\partial f}{\partial y}(x_0, y_0).$$

57.   (a) Set $u = ax + by$. Then

$$b\dfrac{\partial w}{\partial x} - a\dfrac{\partial w}{\partial y} = b(a\, g'(u)) - a(b\, g'(u)) = 0.$$

(b) Set $u = x^m y^n$. Then

$$nx\dfrac{\partial w}{\partial x} - my\dfrac{\partial w}{\partial y} = nx\left[mx^{m-1}y^n g'(u)\right] - my\left[nx^m y^{n-1} g'(u)\right] = 0.$$

59.
$$x\,\frac{\partial u}{\partial x} + y\,\frac{\partial u}{\partial y} = x\left(4Ax^3 + 4Bxy^2\right) + y\left(4Bx^2y + 4Cy^3\right)$$
$$= 4\left(Ax^4 + 2Bx^2y^2 + Cy^4\right) = 4u$$

61. $\dfrac{\partial x}{\partial r}\dfrac{\partial y}{\partial \theta} - \dfrac{\partial x}{\partial \theta}\dfrac{\partial y}{\partial r} = (\cos\theta)(r\cos\theta) - (-r\sin\theta)(\sin\theta) = r$

SECTION 14.5

1. interior $= \{(x,y) : 2 < x < 4,\ 1 < y < 3\}$ (the inside of the rectangle), boundary $=$ the union of the four boundary line segments; set is closed.

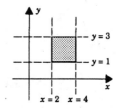

3. interior $=$ the entire set (region between the two concentric circles), boundary $=$ the two circles, one of radius 1, the other of radius 2; set is open.

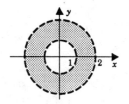

5. interior $= \{(x,y) : 1 < x^2 < 4\} =$
$\{(x,y) : -2 < x < -1\} \cup \{(x,y) : 1 < x < 2\}$
(two vertical strips without the boundary lines),
boundary $= \{(x,y) : x = -2,\ x = -1,\ x = 1,\ \text{or}\ x = 2\}$ (four vertical lines); set is neither open nor closed.

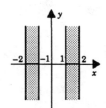

7. interior $=$ region below the parabola $y = x^2$,
boundary $=$ the parabola $y = x^2$; the set is closed.

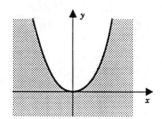

9. interior $= \{(x,y,z) : x^2 + y^2 < 1,\ 0 < z \leq 4\}$
(the inside of the cylinder), boundary $=$ the total surface of the cylinder (the curved part, the top, and the bottom); the set is closed.

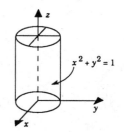

11. (a) ϕ (b) S (c) closed

13. interior $= \{x : 1 < x < 3\}$, boundary $= \{1, 3\}$; set is closed.

15. interior $=$ the entire set, boundary $= \{1\}$; set is open.

17. interior $= \{x : |x| > 1\}$, boundary $= \{1, -1\}$; set is neither open nor closed.

19. interior $= \phi$, boundary $= \{\text{the entire set}\} \cup \{0\}$; the set is neither open nor closed.

SECTION 14.6

1. $\dfrac{\partial^2 f}{\partial x^2} = 2A, \quad \dfrac{\partial^2 f}{\partial y^2} = 2C, \quad \dfrac{\partial^2 f}{\partial y \partial x} = \dfrac{\partial^2 f}{\partial x \partial y} = 2B$

3. $\dfrac{\partial^2 f}{\partial x^2} = Cy^2 e^{xy}, \quad \dfrac{\partial^2 f}{\partial y^2} = Cx^2 e^{xy}, \quad \dfrac{\partial^2 f}{\partial y \partial x} = \dfrac{\partial^2 f}{\partial x \partial y} = Ce^{xy}(xy + 1)$

5. $\dfrac{\partial^2 f}{\partial x^2} = 2, \quad \dfrac{\partial^2 f}{\partial y^2} = 4(x + 3y^2 + z^3), \quad \dfrac{\partial^2 f}{\partial z^2} = 6z(2x + 2y^2 + 5z^3)$

$\dfrac{\partial^2 f}{\partial x \partial y} = \dfrac{\partial^2 f}{\partial y \partial x} = 4y, \quad \dfrac{\partial^2 f}{\partial z \partial x} = \dfrac{\partial^2 f}{\partial x \partial z} = 6z^2, \quad \dfrac{\partial^2 f}{\partial z \partial y} = \dfrac{\partial^2 f}{\partial y \partial z} = 12yz^2$

7. $\dfrac{\partial^2 f}{\partial x^2} = \dfrac{1}{(x + y)^2} - \dfrac{1}{x^2}, \quad \dfrac{\partial^2 f}{\partial y^2} = \dfrac{1}{(x + y)^2}, \quad \dfrac{\partial^2 f}{\partial y \partial x} = \dfrac{\partial^2 f}{\partial x \partial y} = \dfrac{1}{(x + y)^2}$

9. $\dfrac{\partial^2 f}{\partial x^2} = 2(y + z), \quad \dfrac{\partial^2 f}{\partial y^2} = 2(x + z), \quad \dfrac{\partial^2 f}{\partial z^2} = 2(x + y)$

all the second mixed partials are $2(x + y + z)$

11. $\dfrac{\partial^2 f}{\partial x^2} = y(y - 1)x^{y-2}, \quad \dfrac{\partial^2 f}{\partial y^2} = (\ln x)^2 x^y, \quad \dfrac{\partial^2 f}{\partial y \partial x} = \dfrac{\partial^2 f}{\partial x \partial y} = x^{y-1}(1 + y \ln x)$

13. $\dfrac{\partial^2 f}{\partial x^2} = ye^x, \quad \dfrac{\partial^2 f}{\partial y^2} = xe^y, \quad \dfrac{\partial^2 f}{\partial y \partial x} = \dfrac{\partial^2 f}{\partial x \partial y} = e^y + e^x$

15. $\dfrac{\partial^2 f}{\partial x^2} = \dfrac{y^2 - x^2}{(x^2 + y^2)^2}, \quad \dfrac{\partial^2 f}{\partial y^2} = \dfrac{x^2 - y^2}{(x^2 + y^2)^2}, \quad \dfrac{\partial^2 f}{\partial y \partial x} = \dfrac{\partial^2 f}{\partial x \partial y} = -\dfrac{2xy}{(x^2 + y^2)^2}$

17. $\dfrac{\partial^2 f}{\partial x^2} = -2y^2 \cos 2xy, \quad \dfrac{\partial^2 f}{\partial y^2} = -2x^2 \cos 2xy, \quad \dfrac{\partial^2 f}{\partial y \partial x} = \dfrac{\partial^2 f}{\partial x \partial y} = -[\sin 2xy + 2xy \cos 2xy]$

19. $\dfrac{\partial^2 f}{\partial x^2} = 0, \quad \dfrac{\partial^2 f}{\partial y^2} = xz \sin y, \quad \dfrac{\partial^2 f}{\partial z^2} = -xy \sin z,$

$\dfrac{\partial^2 f}{\partial y \partial x} = \dfrac{\partial^2 f}{\partial x \partial y} = \sin z - z \cos y, \quad \dfrac{\partial^2 f}{\partial x \partial z} = \dfrac{\partial^2 f}{\partial z \partial x} = y \cos z - \sin y, \quad \dfrac{\partial^2 f}{\partial y \partial z} = \dfrac{\partial^2 u}{\partial z \partial y} = x \cos z - x \cos y$

21. $x^2 \dfrac{\partial^2 u}{\partial x^2} + 2xy \dfrac{\partial^2 u}{\partial x \partial y} + y^2 \dfrac{\partial^2 u}{\partial y^2} = x^2 \left(\dfrac{-2y^2}{(x+y)^3} \right) + 2xy \left(\dfrac{2xy}{(x+y)^3} \right) + y^2 \left(\dfrac{-2x^2}{(x+y)^3} \right) = 0$

23. $\dfrac{\partial^2 f}{\partial x^2} = e^x \sin y, \quad \dfrac{\partial^2 f}{\partial y^2} = -e^x \sin y, \quad \dfrac{\partial^2 f}{\partial x^2} + \dfrac{\partial^2 f}{\partial y^2} = e^x \sin y + (-e^x \sin y) = 0$

25. $\dfrac{\partial^2 f}{\partial x^2} = \dfrac{y^2 - x^2}{(x^2 + y^2)^2}, \quad \dfrac{\partial^2 f}{\partial y^2} = \dfrac{x^2 - y^2}{(x^2 + y^2)^2}$

$\dfrac{\partial^2 f}{\partial x^2} + \dfrac{\partial^2 f}{\partial y^2} = \dfrac{y^2 - x^2}{(x^2 + y^2)^2} + \dfrac{x^2 - y^2}{(x^2 + y^2)^2} = 0$

27. $\dfrac{\partial^2 f}{\partial x^2} = e^{x+y} \cos\left(\sqrt{2}\, z\right), \quad \dfrac{\partial^2 f}{\partial y^2} = e^{x+y} \cos\left(\sqrt{2}\, z\right), \quad \dfrac{\partial^2 f}{\partial z^2} = -2e^{x+y} \cos\left(\sqrt{2}\, z\right)$

$\dfrac{\partial^2 f}{\partial x^2} + \dfrac{\partial^2 f}{\partial y^2} + \dfrac{\partial^2 f}{\partial z^2} = e^{x+y} \cos\left(\sqrt{2}\, z\right) + e^{x+y} \cos\left(\sqrt{2}\, z\right) + \left[-2e^{x+y} \cos\left(\sqrt{2}\, z\right) \right] = 0$

29. $\dfrac{\partial^2 f}{\partial t^2} = -5c^2 \sin(x+ct)\cos(2x+2ct) - 4c^2 \cos(x+ct)\sin(2x+2ct)$

$\dfrac{\partial^2 f}{\partial x^2} = -5\sin(x+ct)\cos(2x+2ct) - 4\cos(x+ct)\sin(2x+2ct)$

It now follows that $\dfrac{\partial^2 f}{\partial t^2} - c^2 \dfrac{\partial^2 f}{\partial x^2} = 0$

31. $\dfrac{\partial^2 f}{\partial t^2} = c^2 k^2 \left(Ae^{kx} + Be^{-kx}\right)\left(Ce^{ckt} + De^{-ckt}\right), \quad \dfrac{\partial^2 f}{\partial x^2} = k^2 \left(Ae^{kx} + Be^{-kx}\right)\left(Ce^{ckt} + De^{-ckt}\right)$

It now follows that $\dfrac{\partial^2 f}{\partial t^2} - c^2 \dfrac{\partial^2 f}{\partial x^2} = 0$

33. (a) mixed partials are 0

    (b) mixed partials are $g'(x)\, h'(y)$

    (c) by the hint mixed partials for each term $x^m y^n$ are $mn x^{m-1} y^{n-1}$

35. (a) no, since $\dfrac{\partial^2 f}{\partial y \partial x} \neq \dfrac{\partial^2 f}{\partial x \partial y}$ (b) no, since $\dfrac{\partial^2 f}{\partial y \partial x} \neq \dfrac{\partial^2 f}{\partial x \partial y}$ for $x \neq y$

37. $\dfrac{\partial^3 f}{\partial x^2 \partial y} = \dfrac{\partial}{\partial x}\left(\dfrac{\partial^2 f}{\partial x \partial y} \right) = \dfrac{\partial}{\partial x}\left(\dfrac{\partial^2 f}{\partial y \partial x} \right) = \dfrac{\partial^2 f}{\partial x \partial y}\left(\dfrac{\partial f}{\partial x} \right) = \dfrac{\partial^2}{\partial y \partial x}\left(\dfrac{\partial f}{\partial x} \right) = \dfrac{\partial}{\partial y}\left(\dfrac{\partial^2 f}{\partial x^2} \right) = \dfrac{\partial^3 f}{\partial y \partial x^2}$

    by definition — (14.6.5) — by definition — (14.6.5) — by def. — by def.

39. (a) $\lim\limits_{x \to 0} \dfrac{(x)(0)}{x^2 + 0} = \lim\limits_{x \to 0} 0 = 0$ (b) $\lim\limits_{y \to 0} \dfrac{(0)(y)}{0 + y^2} = \lim\limits_{y \to 0} 0 = 0$

    (c) $\lim\limits_{x \to 0} \dfrac{(x)(mx)}{x^2 + (mx)^2} = \lim\limits_{x \to 0} \dfrac{m}{1 + m^2} = \dfrac{m}{1 + m^2}$

(d) $\lim\limits_{\theta\to 0+} \dfrac{(\theta\,\cos\theta)(\theta\,\sin\theta)}{(\theta\,\cos\theta)^2+(\theta\,\sin\theta)^2} = \lim\limits_{\theta\to 0+}\cos\theta\,\sin\theta = 0$

(e)   By L'Hospital's rule $\lim\limits_{x\to 0}\dfrac{f(x)}{x} = \lim\limits_{x\to 0}f'(x) = f'(0)$.   Thus

$$\lim\limits_{x\to 0}\frac{xf(x)}{x^2+[\,f(x)\,]^2} = \lim\limits_{x\to 0}\frac{f(x)/x}{1+[\,f(x)/x\,]^2} = \frac{f'(0)}{1+[\,f'(0)\,]^2}.$$

(f) $\lim\limits_{\theta\to(\pi/3)-} = \dfrac{(\cos\theta\,\sin 3\theta)(\sin\theta\,\sin 3\theta)}{(\cos\theta\,\sin 3\theta)^2+(\sin\theta\,\sin 3\theta)^2} = \lim\limits_{\theta\to(\pi/3)-}\cos\theta\,\sin\theta = \dfrac{1}{4}\sqrt{3}$

(g) $\lim\limits_{t\to\infty}\dfrac{(1/t)(\sin t)/t}{1/t^2+(\sin^2 t)\,/t^2} = \lim\limits_{t\to\infty}\dfrac{\sin t}{1+\sin^2 t}$;   does not exist

41.   (a) $\dfrac{\partial g}{\partial x}(0,0) = \lim\limits_{h\to 0}\dfrac{g(h,0)-g(0,0)}{h} = \lim\limits_{h\to 0}0 = 0,$

$\dfrac{\partial g}{\partial y}(0,0) = \lim\limits_{h\to 0}\dfrac{g(0,h)-g(0,0)}{h} = \lim\limits_{h\to 0}0 = 0$

(b)   as (x,y) tends to $(0,0)$ along the x-axis, $g(x,y) = g(x,0) = 0$ tends to 0;

as (x,y) tends to $(0,0)$ along the line $y=x$, $g(x,y) = g(x,x) = \frac{1}{2}$ tends to $\frac{1}{2}$

43.   For $y\neq 0$, $\dfrac{\partial f}{\partial x}(0,y) = \lim\limits_{h\to 0}\dfrac{f(h,y)-f(0,y)}{h} = \lim\limits_{h\to 0}\dfrac{y(y^2-h^2)}{h^2+y^2} = y.$

Since $\dfrac{\partial f}{\partial x}(0,0) = \lim\limits_{h\to 0}\dfrac{f(h,0)-f(0,0)}{h} = \lim\limits_{h\to 0}0 = 0,$

we have $\dfrac{\partial f}{\partial x}(0,y) = y$ for all y.

For $x\neq 0$, $\dfrac{\partial f}{\partial y}(x,0) = \lim\limits_{h\to 0}\dfrac{f(x,h)-f(x,0)}{h} = \lim\limits_{h\to 0}\dfrac{x(h^2-x^2)}{x^2+h^2} = -x.$

Since $\dfrac{\partial f}{\partial y}(0,0) = \lim\limits_{h\to 0}\dfrac{f(0,h)-f(0,0)}{h} = \lim\limits_{h\to 0}0 = 0,$

we have $\dfrac{\partial f}{\partial y}(x,0) = -x$ for all x.

Therefore $\dfrac{\partial^2 f}{\partial y\partial x}(0,y) = 1$ for all y and $\dfrac{\partial^2 f}{\partial x\partial y}(x,0) = -1$ for all x.

In particular $\dfrac{\partial^2 f}{\partial y\partial x}(0,0) = 1$ while $\dfrac{\partial^2 f}{\partial x\partial y}(0,0) = -1.$

PROJECTS AND EXPLORATIONS

14.1.   (a) Use isosceles triangles rather than equilateral! The base and altitude of each triangle are:

$$b = 2r \, \sin(\pi/m) \quad \text{and} \quad h = \sqrt{\left[\frac{h}{n}\right]^2 + [r - r \, \cos(\pi/m)]^2}.$$

There are $2mn$ such triangles and so the sum of their areas is:

$$2mn \left(\frac{1}{2}\right) [2r \, \sin(\pi/m)] \sqrt{\left[\frac{h}{n}\right]^2 + [r - r \, \cos(\pi/m)]^2}.$$

This expression can be written as:

$$H(m,n) = 2\pi r \, \frac{\sin(\pi/m)}{\pi/m} \left[h^2 + \pi^2 r^2 \left(\frac{n^2}{m^2}\right) \left(\frac{1 - \cos(\pi/m)}{\pi/m}\right)^2\right]^{1/2}$$

(b) Fix n. Then $\displaystyle\lim_{m \to \infty} H(m,n) = 2\pi rh$.

(c) Fix m. Then $\displaystyle\lim_{n \to \infty} H(m,n) = \infty$.

(d) Let $n = m$. Then $\displaystyle\lim_{m \to \infty} H(m,m) = 2\pi rh$.

(e) Let $n = m^2$. Then $\displaystyle\lim_{m \to \infty} H(m,m^2) = 2\pi r\sqrt{h^2 + \frac{1}{4}\pi^4 r^2}$.

These limits can be calculated "by hand," that is, without using technology. However, the following graphs (with $r = h = 1$) help to illustrate why we get these limits.

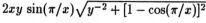

$$2xy \, \sin(\pi/x)\sqrt{y^{-2} + [1 - \cos(\pi/x)]^2}$$

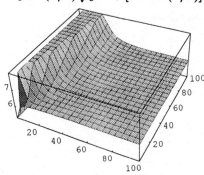

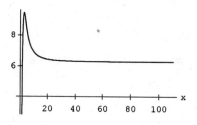

$$2x^2 \, \sin(\pi/x)\sqrt{x^{-2} + [1 - \cos(\pi/x)]^2}$$

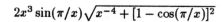

$$2x^3 \, \sin(\pi/x)\sqrt{x^{-4} + [1 - \cos(\pi/x)]^2}$$

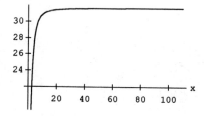

CHAPTER 15

SECTION 15.1

1. $\nabla f = e^{xy}[(xy+1)\mathbf{i} + x^2\mathbf{j}]$

3. $\nabla f = (6x - y)\mathbf{i} + (1 - x)\mathbf{j}$

5. $\nabla f = 2xy^{-2}\mathbf{i} - 2x^2y^{-3}\mathbf{j}$

7. $\nabla f = z\cos(x - y)\mathbf{i} - z\cos(x - y)\mathbf{j} + \sin(x - y)\mathbf{k}$

9. $\nabla f = (y + z)\mathbf{i} + (x + z)\mathbf{j} + (x + y)\mathbf{k}$

11. $\nabla f = e^{x-y}[(1 + x + y)\mathbf{i} + (1 - x - y)\mathbf{j}]$

13. $\nabla f = e^x[\ln y\,\mathbf{i} + y^{-1}\mathbf{j}]$

15. $\nabla f = \dfrac{AD - BC}{(Cx + Dy)^2}[y\mathbf{i} - x\mathbf{j}]$

17. $\nabla f = (ye^x + xye^x - ze^y\cos xz)\,\mathbf{i} + (xe^x + e^z - e^y\sin xz)\,\mathbf{j} + (ye^z - xe^y\cos xz)\,\mathbf{k}$

19. $\nabla f = e^{x+2y}\cos(z^2 + 1)\,\mathbf{i} + 2e^{x+2y}\cos(z^2 + 1)\,\mathbf{j} - 2ze^{x+2y}\sin(z^2 + 1)\,\mathbf{k}$

21. $\nabla f = (4x - 3y)\mathbf{i} + (8y - 3x)\mathbf{j};$ at $(2,3)$, $\nabla f = -\mathbf{i} + 18\mathbf{j}$

23. $\nabla f = \dfrac{2x}{x^2 + y^2}\mathbf{i} + \dfrac{2y}{x^2 + y^2}\mathbf{j};$ at $(2,1)$, $\nabla f = \frac{4}{5}\mathbf{i} + \frac{2}{5}\mathbf{j}$

25. $\nabla f = (\sin xy + xy\cos xy)\mathbf{i} + x^2\cos xy\,\mathbf{j};$ at $(1, \pi/2)$, $\nabla f = \mathbf{i}$

27. $\nabla f = -e^{-x}\sin(z + 2y)\mathbf{i} + 2e^{-x}\cos(z + 2y)\mathbf{j} + e^{-x}\cos(z + 2y)\mathbf{k};$

    at $(0, \pi/4, \pi/4)$, $\nabla f = -\frac{1}{2}\sqrt{2}(\mathbf{i} + 2\mathbf{j} + \mathbf{k})$

29. $\nabla f = \mathbf{i} - \dfrac{y}{\sqrt{y^2 + z^2}}\mathbf{j} - \dfrac{z}{\sqrt{y^2 + z^2}}\mathbf{k};$ at $(2, -3, 4)$, $\nabla f = \mathbf{i} + \frac{3}{5}\mathbf{j} - \frac{4}{5}\mathbf{k}$

31. For the function $f(x, y) = 3x^2 - xy + y$, we have

$$f(\mathbf{x} + \mathbf{h}) - f(\mathbf{x}) = f(x + h_1, y + h_2) - f(x, y)$$

$$= 3(x + h_1)^2 - (x + h_1)(y + h_2) + (y + h_2) - [3x^3 - xy + y]$$

$$= [(6x - y)\mathbf{i} + (1 - x)\mathbf{j}] \cdot (h_1\mathbf{i} + h_2\mathbf{j}) + 3h_1^2 - h_1h_2$$

$$= [(6x - y)\mathbf{i} + (1 - x)\mathbf{j}] \cdot \mathbf{h} + 3h_1^2 - h_1h_2$$

The remainder $g(\mathbf{h}) = 3h_1^2 - h_1h_2 = (3h_1\mathbf{i} - h_1\mathbf{j}) \cdot (h_1\mathbf{i} + h_2\mathbf{j})$,   and

$$\frac{|g(\mathbf{h})|}{\|\mathbf{h}\|} = \frac{\|3h_1\mathbf{i} - h_1\mathbf{j}\| \cdot \|\mathbf{h}\| \cdot \cos\theta}{\|\mathbf{h}\|} \leq \|3h_1\mathbf{i} - h_1\mathbf{j}\|$$

Since $\|3h_1\mathbf{i} - h_1\mathbf{j}\| \to 0$ as $\mathbf{h} \to \mathbf{0}$ it follows that

$$\nabla f = (6x - y)\mathbf{i} + (1 - x)\mathbf{j}$$

33. For the function $f(x,y,z) = x^2y + y^2z + z^2x$, we have

$$f(\mathbf{x}+\mathbf{h}) - f(\mathbf{x}) = f(x+h_1, y+h_2, z+h_3) - f(x,y,z)$$

$$= (x+h_1)^2(y+h_2) + (y+h_2)^2(z+h_3) + (z+h_3)^2(x+h_1) - x^2y + y^2z + z^2x$$

$$= (2xy+z^2)h_1 + (2yz+x^2)h_2 + (2xz+y^2)h_3 + (2xh_2 + yh_1 + h_1h_2)h_1 +$$

$$(2yh_3 + zh_2 + h_2h_3)h_2 + (2zh_1 + xh_3 + h_1h_3)h_3$$

$$= [(2xy+z^2)\mathbf{i} + (2yz+x^2)\mathbf{j} + (2xz+y^2)\mathbf{k}] \cdot \mathbf{h} + g(\mathbf{h}) \cdot \mathbf{h},$$

where $g(\mathbf{h}) = (2xh_2 + yh_1 + h_1h_2)\mathbf{i} + (2yh_3 + zh_2 + h_2h_3)\mathbf{j} + (2zh_1 + xh_3 + h_1h_3)\mathbf{k}$

Since $\dfrac{|g(\mathbf{h})|}{\|\mathbf{h}\|} \to 0$ as $\mathbf{h} \to \mathbf{0}$ it follows that

$$\nabla f = (2xy+z^2)\mathbf{i} + (2yz+x^2)\mathbf{j} + (2xz+y^2)\mathbf{k}$$

35. $\nabla f = \mathbf{F}(x,y) = 2xy\,\mathbf{i} + (1+x^2)\mathbf{j} \Rightarrow \dfrac{\partial f}{\partial x} = 2xy \Rightarrow f(x,y) = x^2y + g(y)$ for some function g.

Now, $\dfrac{\partial f}{\partial y} = x^2 + g'(y) = 1 + x^2 \Rightarrow g'(y) = 1 \Rightarrow g(y) = y + C$, C a constant.

Thus, $f(x,y) = x^2y + y$ (take $C=0$) is a function whose gradient is $\mathbf{F}$.

37. $\nabla f = \mathbf{F}(x,y) = (x+\sin y)\mathbf{i} + (x\cos y - 2y)\mathbf{j} \Rightarrow \dfrac{\partial f}{\partial x} = x+\sin y \Rightarrow f(x,y) = \tfrac{1}{2}x^2 + x\sin y + g(y)$

for some function g.

Now, $\dfrac{\partial f}{\partial y} = x\cos y + g'(y) = x\cos y - 2y \Rightarrow g'(y) = -2y \Rightarrow g(y) = -y^2 + C$, C a constant.

Thus, $f(x,y) = \tfrac{1}{2}x^2 + x\sin y - y^2$ (take $C=0$) is a function whose gradient is $\mathbf{F}$.

39. With $r = (x^2+y^2+z^2)^{1/2}$ we have

$$\frac{\partial r}{\partial x} = \frac{x}{r}, \quad \frac{\partial r}{\partial y} = \frac{y}{r}, \quad \frac{\partial r}{\partial z} = \frac{z}{r}.$$

(a)

$$\nabla(\ln r) = \frac{\partial}{\partial x}(\ln r)\mathbf{i} + \frac{\partial}{\partial y}(\ln r)\mathbf{j} + \frac{\partial}{\partial z}(\ln r)\mathbf{k}$$

$$= \frac{1}{r}\frac{\partial r}{\partial x}\mathbf{i} + \frac{1}{r}\frac{\partial r}{\partial y}\mathbf{j} + \frac{1}{r}\frac{\partial r}{\partial z}\mathbf{k}$$

$$= \frac{x}{r^2}\mathbf{i} + \frac{y}{r^2}\mathbf{j} + \frac{z}{r^2}\mathbf{k}$$

$$= \frac{\mathbf{r}}{r^2}$$

(b)
$$\nabla(\sin r) = \frac{\partial}{\partial x}(\sin r)\mathbf{i} + \frac{\partial}{\partial y}(\sin r)\mathbf{j} + \frac{\partial}{\partial z}(\sin r)\mathbf{k}$$

$$= \cos r \frac{\partial r}{\partial x}\mathbf{i} + \cos r \frac{\partial r}{\partial y}\mathbf{j} + \cos r \frac{\partial r}{\partial z}\mathbf{k}$$

$$= (\cos r)\frac{x}{r}\mathbf{i} + (\cos r)\frac{y}{r}\mathbf{j} + (\cos r)\frac{z}{r}\mathbf{k}$$

$$= \left(\frac{\cos r}{r}\right)\mathbf{r}$$

(c) $\nabla e^r = \left(\dfrac{e^r}{r}\right)\mathbf{r}$ [same method as in (a) and (b)]

41.   (a) $\nabla f = 2x\,\mathbf{i} + 2y\,\mathbf{j} = \mathbf{0} \implies x = y = 0;\quad \nabla f = \mathbf{0}$ at $(0,0)$.

(b)   (c) f has an absolute minimum at $(0,0)$

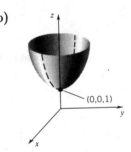

43.   (a) Let $\mathbf{c} = c_1\mathbf{i} + c_2\mathbf{j} + c_3\mathbf{k}$. First, we take $\mathbf{h} = h\mathbf{i}$. Since $\mathbf{c}\cdot\mathbf{h}$ is $o(\mathbf{h})$,

$$0 = \lim_{\mathbf{h}\to 0}\frac{\mathbf{c}\cdot\mathbf{h}}{\|\mathbf{h}\|} = \lim_{h\to 0}\frac{c_1 h}{h} = c_1.$$

Similarly, $c_2 = 0$ and $c_3 = 0$.

(b) $(\mathbf{y} - \mathbf{z})\cdot\mathbf{h} = [f(\mathbf{x}+\mathbf{h}) - f(\mathbf{x}) - \mathbf{z}\cdot\mathbf{h}] + [\mathbf{y}\cdot\mathbf{h} - f(\mathbf{x}+\mathbf{h}) + f(\mathbf{x})] = o(\mathbf{h}) + o(\mathbf{h}) = o(\mathbf{h})$,
so that, by part (a), $\mathbf{y} - \mathbf{z} = \mathbf{0}$.

45.   (a) In Section 14.6 we showed that f was not continuous at $(0,0)$. It is therefore not differentiable at $(0,0)$.

(b) For $(x,y) \ne (0,0)$, $\dfrac{\partial f}{\partial x} = \dfrac{2y(y^2 - x^2)}{(x^2 + y^2)^2}$. As (x,y) tends to $(0,0)$ along the positive y-axis,

$\dfrac{\partial f}{\partial x} = \dfrac{2y^3}{y^4} = \dfrac{2}{y}$ tends to ∞.

47.   (a) $\nabla f = \left(4y - 8x^2 y\right)e^{-(x^2+y^2)}\mathbf{i} + \left(4x - 8xy^2\right)e^{-(x^2+y^2)}\mathbf{j} = \mathbf{0} \implies$

$$4y - 8x^2 y = 0 \quad \text{and} \quad 4x - 8xy^2 = 0$$

The solutions of this system of equations are: $(0,0),\ \left(\pm\dfrac{\sqrt{2}}{2}, \pm\dfrac{\sqrt{2}}{2}\right)$

(b)

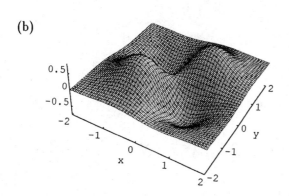

$\left(\dfrac{\sqrt{2}}{2}, \dfrac{\sqrt{2}}{2}\right)$ maximum;

$\left(-\dfrac{\sqrt{2}}{2}, \dfrac{\sqrt{2}}{2}\right)$ minimum;

$\left(\dfrac{\sqrt{2}}{2}, -\dfrac{\sqrt{2}}{2}\right)$ minimum;

$\left(-\dfrac{\sqrt{2}}{2}, -\dfrac{\sqrt{2}}{2}\right)$ maximum.

SECTION 15.2

1. $\nabla f = 2x\mathbf{i} + 6y\mathbf{j}, \quad \nabla f(1,1) = 2\mathbf{i} + 6\mathbf{j}, \quad \mathbf{u} = \tfrac{1}{2}\sqrt{2}\,(\mathbf{i} - \mathbf{j}),$

   $f'_{\mathbf{u}}(1,1) = \nabla f(1,1) \cdot \mathbf{u} = -2\sqrt{2}$

3. $\nabla f\,(e^y - ye^x)\mathbf{i} + (xe^y - e^x)\mathbf{j}, \quad \nabla f(1,0) = \mathbf{i} + (1-e)\mathbf{j}, \quad \mathbf{u} = \tfrac{1}{5}(3\mathbf{i} + 4\mathbf{j}),$

   $f'_{\mathbf{u}}(1,0) = \nabla f(1,0) \cdot \mathbf{u} = \tfrac{1}{5}(7 - 4e)$

5. $\nabla f = \dfrac{(a-b)y}{(x+y)^2}\,\mathbf{i} + \dfrac{(b-a)x}{(x+y)^2}\,\mathbf{j}, \quad \nabla f(1,1) = \dfrac{a-b}{4}(\mathbf{i} - \mathbf{j}), \quad \mathbf{u} = \tfrac{1}{2}\sqrt{2}\,(\mathbf{i} - \mathbf{j}),$

   $f'_{\mathbf{u}}(1,1) = \nabla f(1,1) \cdot \mathbf{u} = \tfrac{1}{4}\sqrt{2}\,(a - b)$

7. $\nabla f = \dfrac{2x}{x^2 + y^2}\,\mathbf{i} + \dfrac{2y}{x^2 + y^2}\,\mathbf{j}, \quad \nabla f(0,1) = 2\mathbf{j}, \quad \mathbf{u} = \dfrac{1}{\sqrt{65}}(8\mathbf{i} + \mathbf{j}),$

   $f'_{\mathbf{u}}(0,1) = \nabla f \cdot \mathbf{u} = \dfrac{2}{\sqrt{65}}$

9. $\nabla f = (y+z)\mathbf{i} + (x+z)\mathbf{j} + (y+x)\mathbf{k}, \quad \nabla f(1,-1,1) = 2\mathbf{j}, \quad \mathbf{u} = \tfrac{1}{6}\sqrt{6}\,(\mathbf{i} + 2\mathbf{j} + \mathbf{k}),$

   $f'_{\mathbf{u}}(1,-1,1) = \nabla f(1,-1,1) \cdot \mathbf{u} = \tfrac{2}{3}\sqrt{6}$

11. $\nabla f = 2\left(x + y^2 + z^2\right)\left(\mathbf{i} + 2y\mathbf{j} + 3z^2\mathbf{k}\right), \quad \nabla f(1,-1,1) = 6(\mathbf{i} - 2\mathbf{j} + 3\mathbf{k}), \quad \mathbf{u} = \tfrac{1}{2}\sqrt{2}\,(\mathbf{i} + \mathbf{j}),$

$f'_{\mathbf{u}}(1,-1,1) = \nabla f(1,-1,1) \cdot \mathbf{u} = -3\sqrt{2}$

13. $\nabla f = \tan^{-1}(y+z)\mathbf{i} + \dfrac{x}{1+(y+z)^2}\mathbf{j} + \dfrac{x}{1+(y+z)^2}\mathbf{k}, \quad \nabla f(1,0,1) = \dfrac{\pi}{4}\mathbf{i} + \dfrac{1}{2}\mathbf{j} + \dfrac{1}{2}\mathbf{k},$

$\mathbf{u} = \dfrac{1}{\sqrt{3}}(\mathbf{i}+\mathbf{j}-\mathbf{k}), \quad f'_{\mathbf{u}}(1,0,1) = \nabla f(1,0,1) \cdot \mathbf{u} = \dfrac{\pi}{4\sqrt{3}} = \dfrac{\sqrt{3}}{12}\pi$

15. $\nabla f = \dfrac{x}{x^2+y^2}\mathbf{i} + \dfrac{y}{x^2+y^2}\mathbf{j}, \quad \mathbf{u} = \dfrac{1}{\sqrt{x^2+y^2}}(-x\mathbf{i}-y\mathbf{j}), \quad f'_{\mathbf{u}}(x,y) = \nabla f \cdot \mathbf{u} = -\dfrac{1}{\sqrt{x^2+y^2}}$

17. $\nabla f = (2Ax + 2By)\mathbf{i} + (2Bx + 2Cy)\mathbf{j}, \quad \nabla f(a,b) = (2aA + 2bB)\mathbf{i} + (2aB + 2bC)\mathbf{j}$

(a) $\mathbf{u} = \tfrac{1}{2}\sqrt{2}(-\mathbf{i}+\mathbf{j}), \quad f'_{\mathbf{u}}(a,b) = \nabla f(a,b) \cdot \mathbf{u} = \sqrt{2}\,[a(B-A) + b(C-B)]$

(b) $\mathbf{u} = \tfrac{1}{2}\sqrt{2}(\mathbf{i}-\mathbf{j}), \quad f'_{\mathbf{u}}(a,b) = \nabla f(a,b) \cdot \mathbf{u} = \sqrt{2}\,[a(A-B) + b(B-C)]$

19. $\nabla f = e^{y^2-z^2}(\mathbf{i} + 2xy\mathbf{j} - 2xz\mathbf{k}), \quad \nabla f(1,2,-2) = \mathbf{i} + 4\mathbf{j} + 4\mathbf{k}, \quad \mathbf{r}'(t) = \mathbf{i} - 2\sin{(t-1)}\mathbf{j} - 2e^{t-1}\mathbf{k},$

at $(1,2,-2) \quad t = 1, \quad \mathbf{r}'(1) = \mathbf{i} - 2\mathbf{k}, \quad \mathbf{u} = \tfrac{1}{5}\sqrt{5}\,(\mathbf{i} - 2\mathbf{k}), \quad f'_{\mathbf{u}}(1,2,-2) = \nabla f(1,2,-2) \cdot \mathbf{u} = -\tfrac{7}{5}\sqrt{5}$

21. $\nabla f = (2x + 2yz)\mathbf{i} + \left(2xz - z^2\right)\mathbf{j} + (2xy - 2yz)\mathbf{k}, \quad \nabla f(1,1,2) = 6\mathbf{i} - 2\mathbf{k}$

The vector $\mathbf{v} = 2\mathbf{i} + \mathbf{j} - 3\mathbf{k}$ is a direction vector for the given line; $\mathbf{u} = \dfrac{1}{\sqrt{14}}(2\mathbf{i} + \mathbf{j} - 3\mathbf{k})$

is a corresponding unit vector; $\quad f'_{\mathbf{u}}(1,1,2) = \nabla f(1,1,2) \cdot \mathbf{u} = \dfrac{18}{\sqrt{14}}$

Note: we could have used $\mathbf{u} = -\dfrac{1}{\sqrt{14}}(2\mathbf{i} + \mathbf{j} - 3\mathbf{k})$ as the unit vector. In this case,

$f'_{\mathbf{u}}(1,1,2) = -\dfrac{18}{\sqrt{14}}$

23. $\nabla f = 2y^2 e^{2x}\mathbf{i} + 2y e^{2x}\mathbf{j}, \quad \nabla f(0,1) = 2\mathbf{i} + 2\mathbf{j}, \quad \|\nabla f\| = 2\sqrt{2}, \quad \dfrac{\nabla f}{\|\nabla f\|} = \dfrac{1}{\sqrt{2}}(\mathbf{i}+\mathbf{j})$

f increases most rapidly in the direction $\mathbf{u} = \dfrac{1}{\sqrt{2}}(\mathbf{i}+\mathbf{j})$; the rate of change is $2\sqrt{2}$.

f decreases most rapidly in the direction $\mathbf{v} = -\dfrac{1}{\sqrt{2}}(\mathbf{i}+\mathbf{j})$; the rate of change is $-2\sqrt{2}$.

25. $\nabla f = \dfrac{x}{\sqrt{x^2+y^2+z^2}}\mathbf{i} + \dfrac{y}{\sqrt{x^2+y^2+z^2}}\mathbf{j} + \dfrac{z}{\sqrt{x^2+y^2+z^2}}\mathbf{k},$

$\nabla f(1,-2,1) = \dfrac{1}{\sqrt{6}}(\mathbf{i} - 2\mathbf{j} + \mathbf{k}), \quad \|\nabla f\| = 1$

f increases most rapidly in the direction $\mathbf{u} = \dfrac{1}{\sqrt{6}}(\mathbf{i} - 2\mathbf{j} + \mathbf{k})$; the rate of change is 1.

f decreases most rapidly in the direction $\mathbf{v} = -\dfrac{1}{\sqrt{6}}(\mathbf{i} - 2\mathbf{j} + \mathbf{k})$; the rate of change is -1.

27. $\nabla f = f'(x_0)\mathbf{i}$. If $f'(x_0) \neq 0$, the gradient points in the direction in which f increases: to the right if $f'(x_0) > 0$, to the left if $f'(x_0) < 0$.

29. (a) $\displaystyle\lim_{h\to 0}\frac{f(h,0)-f(0,0)}{h} = \lim_{h\to 0}\frac{\sqrt{h^2}}{h} = \lim_{h\to 0}\frac{|h|}{h}$ does not exist

(b) no; by Theorem 15.2.5 f cannot be differentiable at $(0,0)$

31. $\nabla\lambda(x,y) = -\frac{8}{3}x\mathbf{i} - 6y\mathbf{j}$

(a) $\nabla\lambda(1,-1) = -\frac{8}{3}\mathbf{i} = 6\mathbf{j}$, $\mathbf{u} = \dfrac{-\nabla\lambda(1,-1)}{\|\nabla\lambda(1,-1)\|} = \dfrac{\frac{8}{3}\mathbf{i}-6\mathbf{j}}{\frac{2}{3}\sqrt{97}}$, $\lambda'_{\mathbf{u}}(1,-1) = \nabla\lambda(1,-1)\cdot\mathbf{u} = -\frac{2}{3}\sqrt{97}$

(b) $\mathbf{u} = \mathbf{i}$, $\lambda'_{\mathbf{u}}(1,2) = \nabla\lambda(1,2)\cdot\mathbf{u} = \left(-\frac{8}{3}\mathbf{i} - 12\mathbf{j}\right)\cdot\mathbf{i} = -\frac{8}{3}$

(c) $\mathbf{u} = \frac{1}{2}\sqrt{2}\,(\mathbf{i}+\mathbf{j})$, $\lambda'_{\mathbf{u}}(2,2) = \nabla\lambda(2,2)\cdot\mathbf{u} = \left(-\frac{16}{3}\mathbf{i} - 12\mathbf{j}\right)\cdot\left[\frac{1}{2}\sqrt{2}\,(\mathbf{i}+\mathbf{j})\right] = -\frac{26}{3}\sqrt{2}$

33. (a) The projection of the path onto the xy-plane is the curve

$$C:\ \mathbf{r}(t) = x(t)\mathbf{i} + y(t)\mathbf{j}$$

which begins at $(1,1)$ and at each point has its tangent vector in the direction of $-\nabla f$. Since

$$\nabla f = 2x\mathbf{i} + 6y\mathbf{j},$$

we have the initial-value problems

$$x'(t) = -2x(t),\quad x(0) = 1 \qquad\text{and}\qquad y'(t) = -6y(t),\quad y(0) = 1.$$

From Theorem 7.6.1 we find that

$$x(t) = e^{-2t} \qquad\text{and}\qquad y(t) = e^{-6t}.$$

Eliminating the parameter t, we find that C is the curve $y = x^3$ from $(1,1)$ to $(0,0)$.

(b) Here

$$x'(t) = -2x(t),\quad x(0) = 1 \qquad\text{and}\qquad y'(t) = -6y(t),\quad y(0) = -2$$

so that

$$x(t) = e^{-2t} \qquad\text{and}\qquad y(t) = -2e^{-6t}.$$

Eliminating the parameter t, we find that the projection of the path onto the xy-plane is the curve $y = -2x^3$ from $(1,-2)$ to $(0,0)$.

35. The projection of the path onto the xy-plane is the curve

$$C:\ \mathbf{r}(t) = x(t)\mathbf{i} + y(t)\mathbf{j}$$

which begins at (a^2, b^2) and at each point has its tangent vector in the direction of $-\nabla f = -\left(2a^2x\mathbf{i} + 2b^2y\mathbf{j}\right)$. Thus,

$$x'(t) = -2a^2 x(t), \quad x(0) = a^2 \qquad \text{and} \qquad y'(t) = -2b^2 y(t), \quad y(0) = b^2$$

so that

$$x(t) = a^2 e^{-2a^2 t} \qquad \text{and} \qquad y(t) = b^2 e^{-2b^2 t}.$$

Since

$$\left[\frac{x}{a^2}\right]^{b^2} = \left(e^{-2a^2 t}\right)^{b^2} = \left[\frac{y}{b^2}\right]^{a^2},$$

C is the curve $(b^2)^{a^2} x^{b^2} = (a^2)^{b^2} y^{a^2}$ from (a^2, b^2) to $(0,0)$.

37. We want the curve

$$C: \ \mathbf{r}(t) = x(t)\mathbf{i} + y(t)\mathbf{j}$$

which begins at $(\pi/4, 0)$ and at each point has its tangent vector in the direction of

$$\nabla T = -\sqrt{2}\, e^{-y} \sin x\, \mathbf{i} - \sqrt{2}\, e^{-y} \cos x\, \mathbf{j}.$$

From

$$x'(t) = -\sqrt{2}\, e^{-y} \sin x \qquad \text{and} \qquad y'(t) = -\sqrt{2}\, e^{-y} \cos x$$

we obtain

$$\frac{dy}{dx} = \frac{y'(t)}{x'(t)} = \cot x$$

so that

$$y = \ln|\sin x| + C.$$

Since $y = 0$ when $x = \pi/4$, we get $C = \ln\sqrt{2}$ and $y = \ln|\sqrt{2} \sin x|$. As $\nabla T(\pi/4, 0) = -\mathbf{i} - \mathbf{j}$, the curve $y = \ln|\sqrt{2}\, \sin x|$ is followed in the direction of decreasing x.

39. (a)
$$\lim_{h \to 0} \frac{f(2+h, (2+h)^2) - f(2,4)}{h} = \lim_{h \to 0} \frac{3(2+h)^2 + (2+h)^2 - 16}{h}$$

$$= \lim_{h \to 0} 4\left[\frac{4h + h^2}{h}\right] = \lim_{h \to 0} 4(4+h) = 16$$

(b)
$$\lim_{h \to 0} \frac{f\left(\frac{h+8}{4}, 4+h\right) - f(2,4)}{h} = \lim_{h \to 0} \frac{3\left(\frac{h+8}{4}\right)^2 + (4+h) - 16}{h}$$

$$= \lim_{h \to 0} \frac{\frac{3}{16}h^2 + 3h + 12 + 4 + h - 16}{h}$$

$$= \lim_{h \to 0} \left(\frac{3}{16}h + 4\right) = 4$$

(c) $\mathbf{u} = \frac{1}{17}\sqrt{17}\,(\mathbf{i} + 4\mathbf{j}), \quad \nabla f(2,4) = 12\mathbf{i} + \mathbf{j}; \quad f'_{\mathbf{u}}(2,4) = \nabla f(2,4) \cdot \mathbf{u} = \frac{16}{17}\sqrt{17}$

(d)   The limits computed in (a) and (b) are not directional derivatives. In (a) and (b) we have, in essence, computed $\nabla f(2,4) \cdot \mathbf{r}_0$ taking $\mathbf{r}_0 = \mathbf{i} + 4\mathbf{j}$ in (a) and $\mathbf{r}_0 = \frac{1}{4}\mathbf{i} + \mathbf{j}$ in (b). In neither case is $\mathbf{r}_0$ a unit vector.

41.   (a) $\mathbf{u} = \cos\theta\,\mathbf{i} + \sin\theta\,\mathbf{j},\qquad \nabla f(x,y) = \dfrac{\partial f}{\partial x}\mathbf{i} + \dfrac{\partial f}{\partial y}\mathbf{j};$

$$f'_{\mathbf{u}}(x,y) = \nabla f \cdot \mathbf{u} = \left(\frac{\partial f}{\partial x}\mathbf{i} + \frac{\partial f}{\partial y}\mathbf{j}\right)\cdot(\cos\theta\,\mathbf{i} + \sin\theta\,\mathbf{j}) = \frac{\partial f}{\partial x}\cos\theta + \frac{\partial f}{\partial y}\sin\theta$$

(b) $\nabla f = \left(3x^2 + 2y - y^2\right)\mathbf{i} + (2x - 2xy)\mathbf{j},\qquad \nabla f(-1,2) = 3\mathbf{i} + 2\mathbf{j}$

$$f'_{\mathbf{u}}(-1,2) = 3\cos(2\pi/3) + 2\sin(2\pi/3) = \frac{2\sqrt{3} - 3}{2}$$

43. $$\nabla(fg) = \frac{\partial fg}{\partial x}\mathbf{i} + \frac{\partial fg}{\partial y}\mathbf{j} = \left(f\frac{\partial g}{\partial x} + g\frac{\partial f}{\partial x}\right)\mathbf{i} + \left(f\frac{\partial g}{\partial y} + g\frac{\partial f}{\partial y}\right)\mathbf{j}$$

$$= f\left(\frac{\partial g}{\partial x}\mathbf{i} + \frac{\partial g}{\partial y}\mathbf{j}\right) + g\left(\frac{\partial f}{\partial x}\mathbf{i} + \frac{\partial f}{\partial y}\mathbf{j}\right) = f\nabla g + g\nabla f$$

45. $$\nabla f^n = \frac{\partial f^n}{\partial x}\mathbf{i} + \frac{\partial f^n}{\partial y}\mathbf{j} = nf^{n-1}\frac{\partial f}{\partial x}\mathbf{i} + nf^{n-1}\frac{\partial f}{\partial y}\mathbf{j} = nf^{n-1}\nabla f$$

SECTION 15.3

1. $f(\mathbf{b}) = f(1,3) = -2;\;\; f(\mathbf{a}) = f(0,1) = 0;\;\; f(\mathbf{b}) - f(\mathbf{a}) = -2$

$\nabla f = \left(3x^2 - y\right)\mathbf{i} - x\mathbf{j};\;\; \mathbf{b} - \mathbf{a} = \mathbf{i} + 2\mathbf{j}$ and $\nabla f \cdot (\mathbf{b} - \mathbf{a}) = 3x^2 - y - 2x$

The line segment joining $\mathbf{a}$ and $\mathbf{b}$ is parametrized by

$$x = t,\quad y = 1 + 2t,\quad 0 \le t \le 1$$

Thus, we need to solve the equation

$$3t^2 - (1 + 2t) - 2t = -2,\quad \text{which is the same as}\quad 3t^2 - 4t + 1 = 0,\;\; 0 \le t \le 1$$

The solutions are: $t = \frac{1}{3}, t = 1$. Thus, $\mathbf{c} = \left(\frac{1}{3}, \frac{5}{3}\right)$ satisfies the equation.

Note that the endpoint $\mathbf{b}$ also satisfies the equation.

3.   (a) $f(x,y,z) = a_1 x + a_2 y + a_3 z + C$ (b) $f(x,y,z) = g(x,y,z) + a_1 x + a_2 y + a_3 z + C$

5.   (a) U is not connected

(b) (i) $g(\mathbf{x}) = f(\mathbf{x}) - 1$ (ii) $g(\mathbf{x}) = -f(\mathbf{x})$

(c) U is not connected

7.   Since f is continuous at $\mathbf{a}$ and $f(\mathbf{a}) = A$, there exists $\delta > 0$ such that

$$\text{if}\quad \|\mathbf{x} - \mathbf{a}\| < \delta\quad \text{and}\quad \mathbf{x} \in \Omega,\quad \text{then}\quad |f(\mathbf{x}) - A| < \epsilon.$$

Whether $\mathbf{a}$ is on the boundary of Ω or in the interior of Ω, there exists $\mathbf{x}_1$ in the interior of Ω within δ of $\mathbf{a}$. That implies that $|f(\mathbf{x}_1) - A| < \epsilon$ and thus that $f(\mathbf{x}_1) < A + \epsilon$. A similar argument shows the existence of $\mathbf{x}_2$ with the desired property.

SECTION 15.4

1. $\nabla f = 2xy\mathbf{i} + x^2\mathbf{j}$;

   $\nabla f(\mathbf{r}(t)) \cdot \mathbf{r}'(t) = \left(2\mathbf{i} + e^{2t}\mathbf{j}\right) \cdot \left(e^t\mathbf{i} - e^{-t}\mathbf{j}\right) = e^t$

3. $\nabla f = \dfrac{-2x}{1 + (y^2 - x^2)^2}\mathbf{i} + \dfrac{2y}{1 + (y^2 - x^2)^2}\mathbf{j}, \quad \nabla f(\mathbf{r}(t)) = \dfrac{-2\sin t}{1 + \cos^2 2t}\mathbf{i} + \dfrac{2\cos t}{1 + \cos^2 2t}\mathbf{j}$

   $\nabla f(\mathbf{r}(t)) \cdot \mathbf{r}'(t) = \left(\dfrac{-2\sin t}{1 + \cos^2 2t}\mathbf{i} + \dfrac{2\cos t}{1 + \cos^2 2t}\mathbf{j}\right) \cdot (\cos t\,\mathbf{i} - \sin t\,\mathbf{j}) = \dfrac{-4\sin t\,\cos t}{1 + \cos^2 2t} = \dfrac{-2\sin 2t}{1 + \cos^2 2t}$

5. $\nabla f = (e^y - ye^{-x})\,\mathbf{i} + (xe^y + e^{-x})\,\mathbf{j}; \quad \nabla f(\mathbf{r}(t)) = (t^t - \ln t)\,\mathbf{i} + \left(t^t \ln t + \dfrac{1}{t}\right)\mathbf{j}$

   $\nabla f(\mathbf{r}(t)) \cdot \mathbf{r}'(t) = \left((t^t - \ln t)\,\mathbf{i} + \left(t^t \ln t + \dfrac{1}{t}\right)\mathbf{j}\right) \cdot \left(\dfrac{1}{t}\mathbf{i} + [1 + \ln t]\mathbf{j}\right) = t^t\left(\dfrac{1}{t} + \ln t + [\ln t]^2\right) + \dfrac{1}{t}$

7. $\nabla f = y\mathbf{i} + (x - z)\mathbf{j} - y\mathbf{k}$;

   $\nabla f(\mathbf{r}(t)) \cdot \mathbf{r}'(t) = \left(t^2\mathbf{i} + (t - t^3)\mathbf{j} - t^2\mathbf{k}\right) \cdot \left(\mathbf{i} + 2t\mathbf{j} + 3t^2\mathbf{k}\right) = 3t^2 - 5t^4$

9. $\nabla f = 2x\mathbf{i} + 2y\mathbf{j} + \mathbf{k}$;

   $\nabla f(\mathbf{r}(t)) \cdot \mathbf{r}'(t) = (2a\cos\omega t\,\mathbf{i} + 2b\sin\omega t\,\mathbf{j} + \mathbf{k}) \cdot (-a\omega\sin\omega t\,\mathbf{i} + b\omega\cos\omega t\,\mathbf{j} + b\omega\mathbf{k})$

   $= 2\omega\left(b^2 - a^2\right)\sin\omega t\,\cos\omega t + b\omega$

11. $\dfrac{du}{dt} = \dfrac{\partial u}{\partial x}\dfrac{dx}{dt} + \dfrac{\partial u}{\partial y}\dfrac{dy}{dt} = (2x - 3y)(-\sin t) + (4y - 3x)(\cos t)$

   $= 2\cos t\,\sin t + 3\sin^2 t - 3\cos^2 t = \sin 2t - 3\cos 2t$

13. $\dfrac{du}{dt} = \dfrac{\partial u}{\partial x}\dfrac{dx}{dt} + \dfrac{\partial u}{\partial y}\dfrac{dy}{dt}$

   $= (e^x \sin y + e^y \cos x)\left(\tfrac{1}{2}\right) + (e^x \cos y + e^y \sin x)(2)$

   $= e^{t/2}\left(\tfrac{1}{2}\sin 2t + 2\cos 2t\right) + e^{2t}\left(\tfrac{1}{2}\cos\tfrac{1}{2}t + 2\sin\tfrac{1}{2}t\right)$

15. $\dfrac{du}{dt} = \dfrac{\partial u}{\partial x}\dfrac{dx}{dt} + \dfrac{\partial u}{\partial y}\dfrac{dy}{dt} = (e^x \sin y)(2t) + (e^x \cos y)(\pi)$

   $= e^{t^2}[2t\,\sin(\pi t) + \pi\,\cos(\pi t)]$

17. $\dfrac{du}{dt} = \dfrac{\partial u}{\partial x}\dfrac{dx}{dt} + \dfrac{\partial u}{\partial y}\dfrac{dy}{dt} + \dfrac{\partial u}{\partial z}\dfrac{dz}{dt}$

$$= (y+z)(2t) + (x+z)(1-2t) + (y+x)(2t-2)$$

$$= (1-t)(2t) + (2t^2 - 2t + 1)(1-2t) + t(2t-2)$$

$$= 1 - 4t + 6t^2 - 4t^3$$

19. $V = \dfrac{1}{3}\pi r^2 h,$ $\dfrac{dV}{dt} = \dfrac{\partial V}{\partial r}\dfrac{dr}{dt} + \dfrac{\partial V}{\partial h}\dfrac{dh}{dt} = \left(\dfrac{2}{3}\pi rh\right)\dfrac{dr}{dt} + \left(\dfrac{1}{3}\pi r^2\right)\dfrac{dh}{dt}.$

At the given instant,

$$\dfrac{dV}{dt} = \dfrac{2}{3}\pi(280)(3) + \dfrac{1}{3}\pi(196)(-2) = \dfrac{1288}{3}\pi.$$

The volume is increasing at the rate of $\frac{1288}{3}\pi$ in.^{3}/ sec .

21. $A = \frac{1}{2}xy\sin\theta;$ $\dfrac{dA}{dt} = \dfrac{\partial A}{\partial x}\dfrac{dx}{dt} + \dfrac{\partial A}{\partial y}\dfrac{dy}{dt} + \dfrac{\partial A}{\partial\theta}\dfrac{d\theta}{dt} = \frac{1}{2}\left[(y\sin\theta)\dfrac{dx}{dt} + (x\sin\theta)\dfrac{dy}{dt} + (xy\cos\theta)\dfrac{d\theta}{dt}\right].$

At the given instant

$$\dfrac{dA}{dt} = \dfrac{1}{2}\left[(2\sin 1)(0.25) + (1.5\sin 1)(0.25) + (2(1.5)\cos 1)(0.1)\right] \cong 0.2871\,\text{ft}^2/s \cong 41.34\,\text{in}^2/s$$

23. $\dfrac{\partial u}{\partial s} = \dfrac{\partial u}{\partial x}\dfrac{\partial x}{\partial s} + \dfrac{\partial u}{\partial y}\dfrac{\partial y}{\partial s} = (2x - y)(\cos t) + (-x)(t\cos s)$

$$= 2s\cos^2 t - t\sin s\cos t - st\cos s\cos t$$

$$\dfrac{\partial u}{\partial t} = \dfrac{\partial u}{\partial x}\dfrac{\partial x}{\partial t} + \dfrac{\partial u}{\partial y}\dfrac{\partial y}{\partial t} = (2x - y)(-s\sin t) + (-x)(\sin s)$$

$$= -2s^2\cos t\sin t + st\sin s\sin t - s\cos t\sin s$$

25. $\dfrac{\partial u}{\partial s} = \dfrac{\partial u}{\partial x}\dfrac{\partial x}{\partial s} + \dfrac{\partial u}{\partial y}\dfrac{\partial y}{\partial s} = (2x\tan y)(2st) + \left(x^2\sec^2 y\right)(1)$

$$= 4s^3 t^2 \tan\left(s + t^2\right) + s^4 t^2 \sec^2\left(s + t^2\right)$$

$$\dfrac{\partial u}{\partial t} = \dfrac{\partial u}{\partial x}\dfrac{\partial x}{\partial t} + \dfrac{\partial u}{\partial y}\dfrac{\partial y}{\partial t = (2x\tan y)\left(s^2\right)} + \left(x^2\sec^2 y\right)(2t)$$

$$= 2s^4 t \tan\left(s + t^2\right) + 2s^4 t^3 \sec^2\left(s + t^2\right)$$

27. $\dfrac{\partial u}{\partial s} = \dfrac{\partial u}{\partial x}\dfrac{\partial x}{\partial s} + \dfrac{\partial u}{\partial y}\dfrac{\partial y}{\partial s} + \dfrac{\partial u}{\partial z}\dfrac{\partial z}{\partial s}$

$$= (2x - y)(\cos t) + (-x)(-\cos(t - s)) + 2z(t\cos s)$$

$$= 2s\cos^2 t - \sin(t - s)\cos t + s\cos t\cos(t - s) + 2t^2\sin s\cos s$$

$$\dfrac{\partial u}{\partial t} = \dfrac{\partial u}{\partial x}\dfrac{\partial x}{\partial t} + \dfrac{\partial u}{\partial y}\dfrac{\partial y}{\partial t} + \dfrac{\partial u}{\partial z}\dfrac{\partial z}{\partial t}$$

$$= (2x - y)(-s\sin t) + (-x)(\cos(t - s)) + 2z(\sin s)$$

$$= -2s^2 \cos t \sin t + s \sin (t - s) \sin t - s \cos t \cos (t - s) + 2t \sin^2 s$$

29.

$$\frac{d}{dt}\left[f(\mathbf{r}(t))\right] = \left[\nabla f(\mathbf{r}(t)) \cdot \frac{\mathbf{r}'(t)}{\|\mathbf{r}'(t)\|}\right] \|\mathbf{r}'(t)\|$$

$$= f'_{\mathbf{u}(t)}(\mathbf{r}(t)) \|\mathbf{r}'(t)\| \quad \text{where} \quad \mathbf{u}(t) = \frac{\mathbf{r}'(t)}{\|\mathbf{r}'(t)\|}$$

31. (a) $(\cos r)\dfrac{\mathbf{r}}{r}$ (b) $(r \cos r + \sin r)\dfrac{\mathbf{r}}{r}$

33. (a) $(r \cos r - \sin r)\dfrac{\mathbf{r}}{r^3}$ (b) $\left(\dfrac{\sin r - r \cos r}{\sin^2 r}\right)\dfrac{\mathbf{r}}{r}$

35. (a)

(b) $\dfrac{\partial u}{\partial r} = \dfrac{\partial u}{\partial x}\left(\dfrac{\partial x}{\partial w}\dfrac{\partial w}{\partial r} + \dfrac{\partial x}{\partial t}\dfrac{\partial t}{\partial r}\right) + \dfrac{\partial u}{\partial y}\left(\dfrac{\partial y}{\partial w}\dfrac{\partial w}{\partial r} + \dfrac{\partial y}{\partial t}\dfrac{\partial t}{\partial r}\right) + \dfrac{\partial u}{\partial z}\left(\dfrac{\partial z}{\partial w}\dfrac{\partial w}{\partial r} + \dfrac{\partial z}{\partial t}\dfrac{\partial t}{\partial r}\right).$

To obtain $\partial u/\partial s$, replace each r by s.

37. $\dfrac{du}{dt} = \dfrac{\partial u}{\partial x}\dfrac{dx}{dt} + \dfrac{\partial u}{\partial y}\dfrac{dy}{dt}$

$$\frac{d^2 u}{dt^2} = \frac{\partial u}{\partial x}\frac{d^2 x}{dt^2} + \frac{dx}{dt}\left[\frac{\partial^2 u}{\partial x^2}\frac{dx}{dt} + \frac{\partial^2 u}{\partial y \partial x}\frac{dy}{dt}\right] + \frac{\partial u}{\partial y}\frac{d^2 y}{dt^2} + \frac{dy}{dt}\left[\frac{\partial^2 u}{\partial x \partial y}\frac{dx}{dt} + \frac{\partial^2 u}{\partial y^2}\frac{dy}{dt}\right]$$

and the result follows.

39. (a) $\dfrac{\partial u}{\partial r} = \dfrac{\partial u}{\partial x}\dfrac{\partial x}{\partial r} + \dfrac{\partial u}{\partial y}\dfrac{\partial y}{\partial r} = \dfrac{\partial u}{\partial x}\cos\theta + \dfrac{\partial u}{\partial y}\sin\theta$

$\dfrac{\partial u}{\partial \theta} = \dfrac{\partial u}{\partial x}\dfrac{\partial x}{\partial \theta} + \dfrac{\partial u}{\partial y}\dfrac{\partial y}{\partial \theta} = \dfrac{\partial u}{\partial x}(-r \sin\theta) + \dfrac{\partial u}{\partial y}(r \cos\theta)$

(b) $\left(\dfrac{\partial u}{\partial r}\right)^2 = \left(\dfrac{\partial u}{\partial x}\right)^2 \cos^2\theta + 2\dfrac{\partial u}{\partial x}\dfrac{\partial u}{\partial y}\cos\theta \sin\theta + \left(\dfrac{\partial u}{\partial y}\right)^2 \sin^2\theta,$

$\dfrac{1}{r^2}\left(\dfrac{\partial u}{\partial \theta}\right)^2 = \left(\dfrac{\partial u}{\partial x}\right)^2 \sin^2\theta - 2\dfrac{\partial u}{\partial x}\dfrac{\partial u}{\partial y}\cos\theta \sin\theta + \left(\dfrac{\partial u}{\partial y}\right)^2 \cos^2\theta,$

$$\left(\frac{\partial u}{\partial r}\right)^2 + \frac{1}{r^2}\left(\frac{\partial u}{\partial \theta}\right)^2 = \left(\frac{\partial u}{\partial x}\right)^2(\cos^2\theta + \sin^2\theta) + \left(\frac{\partial u}{\partial y}\right)^2(\sin^2\theta + \cos^2\theta) = \left(\frac{\partial u}{\partial x}\right)^2 + \left(\frac{\partial u}{\partial y}\right)^2$$

41. Solve the equations in Exercise 39 (a) for $\dfrac{\partial u}{\partial x}$ and $\dfrac{\partial u}{\partial y}$:

$$\frac{\partial u}{\partial x} = \frac{\partial u}{\partial r}\cos\theta - \frac{1}{r}\frac{\partial u}{\partial \theta}\sin\theta, \quad \frac{\partial u}{\partial y} = \frac{\partial u}{\partial r}\sin\theta + \frac{1}{r}\frac{\partial u}{\partial \theta}\cos\theta$$

Then $\quad \nabla u = \dfrac{\partial u}{\partial x}\mathbf{i} + \dfrac{\partial u}{\partial y}\mathbf{j} = \dfrac{\partial u}{\partial r}(\cos\theta\,\mathbf{i} + \sin\theta\,\mathbf{j}) + \dfrac{1}{r}\dfrac{\partial u}{\partial \theta}(-\sin\theta\,\mathbf{i} + \cos\theta\,\mathbf{j})$

43. $u(x,y) = x^2 - xy + y^2 = r^2 - r^2\cos\theta\sin\theta = r^2\left(1 - \frac{1}{2}\sin 2\theta\right)$

$$\frac{\partial u}{\partial r} = r(2 - \sin 2\theta), \quad \frac{\partial u}{\partial \theta} = -r^2\cos 2\theta$$

$$\nabla u = \frac{\partial u}{\partial r}\mathbf{e_r} + \frac{1}{r}\frac{\partial u}{\partial \theta}\mathbf{e_\theta} = r(2 - \sin 2\theta)\mathbf{e_r} - r\cos 2\theta\,\mathbf{e_\theta}$$

45. From Exercise 39 (a),

$$\frac{\partial^2 u}{\partial r^2} = \frac{\partial^2 u}{\partial x^2}\cos^2\theta + 2\frac{\partial^2 u}{\partial y\,\partial x}\sin\theta\,\cos\theta + \frac{\partial^2 u}{\partial y^2}\sin^2\theta$$

$$\frac{\partial^2 u}{\partial \theta^2} = \frac{\partial^2 u}{\partial x^2}r^2\sin^2\theta - 2\frac{\partial^2 u}{\partial y\,\partial x}r^2\sin\theta\,\cos\theta + \frac{\partial^2 u}{\partial y^2}r^2\cos^2\theta - r\left(\frac{\partial u}{\partial x}\cos\theta + \frac{\partial u}{\partial y}\sin\theta\right).$$

The term in parentheses is $\dfrac{\partial u}{\partial r}$. Now divide the second equation by r^2 and add the two equations. The result follows.

47. Set $u = xe^y + ye^x - 2x^2y$. Then

$$\frac{\partial u}{\partial x} = e^y + ye^x - 4xy, \qquad \frac{\partial u}{\partial y} = xe^y + e^x - 2x^2$$

$$\frac{dy}{dx} = -\frac{\partial u/\partial x}{\partial u/\partial y} = -\frac{e^y + ye^x - 4xy}{xe^y + e^x - 2x^2}.$$

49. Set $u = x\cos xy + y\cos x - 2$. Then

$$\frac{\partial u}{\partial x} = \cos xy - xy\sin xy - y\sin x, \qquad \frac{\partial u}{\partial y} = -x^2\sin xy + \cos x$$

$$\frac{dy}{dx} = -\frac{\partial u/\partial x}{\partial u/\partial y} = \frac{\cos xy - xy\sin xy - y\sin x}{x^2\sin xy - \cos x}.$$

51.   Set $u = \cos xyz + \ln \left(x^2 + y^2 + z^2 \right)$.   Then

$$\frac{\partial u}{\partial x} = -yz \sin xyz + \frac{2x}{x^2 + y^2 + z^2}, \quad \frac{\partial u}{\partial y} = -xz \sin xyz + \frac{2y}{x^2 + y^2 + z^2}, \quad \text{and}$$

$$\frac{\partial u}{\partial z} = -xy \sin xyz + \frac{2z}{x^2 + y^2 + z^2}.$$

$$\frac{\partial z}{\partial x} = -\frac{\partial u/\partial x}{\partial u/\partial z} = -\frac{2x - yz \left(x^2 + y^2 + z^2 \right) \sin xyz}{2z - xy \left(x^2 + y^2 + z^2 \right) \sin xyz},$$

$$\frac{\partial z}{\partial y} = -\frac{\partial u/\partial y}{\partial u/\partial z} = -\frac{2y - xz \left(x^2 + y^2 + z^2 \right) \sin xyz}{2z - xy \left(x^2 + y^2 + z^2 \right) \sin xyz}.$$

53.   Set $u = z \tan x - xy^2 z^3 - 2xyz$.   Then

$$\frac{\partial u}{\partial x} = z \sec^2 x - y^2 z^3 - 2yz, \quad \frac{\partial u}{\partial y} = -2xyz^3 - 2xz, \quad \frac{\partial u}{\partial z} = \tan x - 3xy^2 z^2 - 2xy.$$

$$\frac{\partial z}{\partial x} = -\frac{\partial u/\partial x}{\partial u/\partial z} = -\frac{z \sec^2 x - y^2 z^3 - 2yz}{\tan x - 3xy^2 z^2 - 2xy}, \quad \frac{\partial z}{\partial y} = -\frac{\partial u/\partial y}{\partial u/\partial z} = \frac{2xyz^3 + 2xz}{\tan x - 3xy^2 z^2 - 2xy}.$$

55. $\dfrac{\partial \mathbf{u}}{\partial s} = \dfrac{\partial \mathbf{u}}{\partial x} \dfrac{\partial x}{\partial s} + \dfrac{\partial \mathbf{u}}{\partial y} \dfrac{\partial y}{\partial s}, \quad \dfrac{\partial \mathbf{u}}{\partial t} = \dfrac{\partial \mathbf{u}}{\partial x} \dfrac{\partial x}{\partial t} + \dfrac{\partial \mathbf{u}}{\partial y} \dfrac{\partial y}{\partial t}$

57.   (a) $f(x, y) = f\left(x, x\dfrac{y}{x} \right) = x^n f(1, v)$,   where $v = y/x$.   Now differentiate with respect to x :

$$\frac{\partial f}{\partial x} = nx^{n-1} f(1, v) + x^n \frac{df}{dv} \left(-\frac{y}{x^2} \right)$$

Multiplying through by x and noting that $x^{n-1} \dfrac{df}{dv} = \dfrac{\partial f}{\partial y}$,   we get

$$x \frac{\partial f}{\partial x} = nx^{n-1} f(1, v) - yx^{n-1} \frac{df}{dv} = x f(x, y) - y \frac{\partial f}{\partial y}$$

It now follows that

(*) $x \dfrac{\partial f}{\partial x} + y \dfrac{\partial f}{\partial y} = n\, f(x, y)$

(b) Differentiate (*) with respect to x and multiply the resulting equation by x; then differentiate
(*) by y and multiply the resulting equation by y; add the two equations.

SECTION 15.5

1. Set $f(x, y) = x^2 + xy + y^2$. Then,

$$\nabla f = (2x + y)\mathbf{i} + (x + 2y)\mathbf{j}, \quad \nabla f(-1, -1) = -3\mathbf{i} - 3\mathbf{j}.$$

normal vector $\mathbf{i} + \mathbf{j}$; tangent vector $\mathbf{i} - \mathbf{j}$

tangent line $x + y + 2 = 0$; normal line $x - y = 0$

3. Set $f(x, y) = \left(x^2 + y^2\right)^2 - 9\left(x^2 - y^2\right)$. Then,

$$\nabla f = [4x(x^2 + y^2) - 18x]\mathbf{i} + [4y\left(x^2 + y^2\right) + 18y]\mathbf{j}, \quad \nabla f\left(\sqrt{2}, 1\right) = -6\sqrt{2}\,\mathbf{i} + 30\mathbf{j}.$$

normal vector $\sqrt{2}\,\mathbf{i} - 5\mathbf{j}$; tangent vector $5\mathbf{i} + \sqrt{2}\,\mathbf{j}$

tangent line $\sqrt{2}x - 5y + 3 = 0$; normal line $5x + \sqrt{2}\,y - 6\sqrt{2} = 0$

5. Set $f(x, y) = xy^2 - 2x^2 + y + 5x$. Then,

$$\nabla f = (y^2 - 4x + 5)\,\mathbf{i} + (2xy + 1)\mathbf{j}, \quad \nabla f(4, 2) = -7\mathbf{i} + 17\mathbf{j}.$$

normal vector $7\mathbf{i} - 17\mathbf{j}$; tangent vector $17\mathbf{i} + 7\mathbf{j}$

tangent line $7x - 17y + 6 = 0$; normal line $17x + 7y - 82 = 0$

7. Set $f(x, y) = 2x^3 - x^2y^2 - 3x + y$. Then,

$$\nabla f = (6x^2 - 2xy^2 - 3)\,\mathbf{i} + (-2x^2y + 1)\mathbf{j}, \quad \nabla f(1, -2) = -5\mathbf{i} + 5\mathbf{j}.$$

normal vector $\mathbf{i} - \mathbf{j}$; tangent vector $\mathbf{i} + \mathbf{j}$

tangent line $x - y - 3 = 0$; normal line $x + y + 1 = 0$

9. Set $f(x, y) = x^2y + a^2y$. By (15.5.4)

$$m = -\frac{\partial f/\partial x}{\partial f/\partial y} = -\frac{2xy}{x^2 + a^2}.$$

At $(0, a)$ the slope is 0.

11. Set $f(x, y, z) = x^3 + y^3 - 3xyz$. Then,

$$\nabla f = (3x^2 - 3yz)\,\mathbf{i} + (3y^2 - 3xz)\mathbf{j} - 3xy\mathbf{k}, \quad \nabla f\left(1, 2, \tfrac{3}{2}\right) = -6\mathbf{i} + \tfrac{15}{2}\mathbf{j} - 6\mathbf{k};$$

tangent plane at $\left(1, 2, \tfrac{3}{2}\right)$: $-6(x-1) + \tfrac{15}{2}(y-2) - 6\left(z - \tfrac{3}{2}\right) = 0$, which reduces to $4x - 5y + 4z = 0$.

Normal: $x = 1 + 4t$, $y = 2 - 5t$, $z = \tfrac{3}{2} + 4t$

13. Set $z = g(x, y) = axy$. Then, $\nabla g = ay\mathbf{i} + ax\mathbf{j}$, $\nabla g\left(1, \dfrac{1}{a}\right) = \mathbf{i} + a\mathbf{j}$.

tangent plane at $\left(1, \frac{1}{a}, 1\right)$: $z - 1 = 1(x - 1) + a\left(y - \frac{1}{a}\right)$, which reduces to $x + ay - z - 1 = 0$

Normal: $x = 1 + t$, $y = \frac{1}{a} + at$, $z = 1 - t$

15. Set $z = g(x, y) = \sin x + \sin y + \sin(x + y)$.   Then,

$$\nabla g = [\cos x + \cos(x + y)]\mathbf{i} + [\cos y + \cos(x + y)]\mathbf{j}, \quad \nabla g(0,0) = 2\mathbf{i} + 2\mathbf{j};$$

tangent plane at $(0,0,0)$: $z - 0 = 2(x - 0) + 2(y - 0)$, $2x + 2y - z = 0$.

Normal: $x = 2t$, $y = 2t$, $z = -t$

17. Set $f(x, y, z) = b^2 c^2 x^2 - a^2 c^2 y^2 - a^2 b^2 z^2$.   Then,

$$\nabla f(x_0, y_0, z_0) = 2b^2 c^2 x_0 \mathbf{i} - 2a^2 c^2 y_0 \mathbf{j} - 2a^2 b^2 z_0 \mathbf{k};$$

tangent plane at (x_0, y_0, z_0): $2b^2 c^2 x_0 (x - x_0) - 2a^2 c^2 y_0 (y - y_0) - 2a^2 b^2 z_0 (z - z_0) = 0$,

which can be rewritten as: $b^2 c^2 x_0 x - a^2 c^2 y_0 y - a^2 b^2 z_0 z = a^2 + b^2 + c^2$.

Normal: $x = x_0 + b^2 c^2 x_0 t$, $y = y_0 - a^2 c^2 y_0 t$, $z = z_0 - a^2 b^2 z_0 t$

19. Set $z = g(x, y) = xy + a^3 x^{-1} + b^3 y^{-1}$.

$$\nabla g = \left(y - a^3 x^{-2}\right)\mathbf{i} + \left(x - b^3 y^{-2}\right)\mathbf{j}, \quad \nabla g = 0 \implies y = a^3 x^{-2} \text{ and } x = b^3 y^{-2}.$$

Thus,

$$y = a^3 b^{-6} y^4, \quad y^3 = b^6 a^{-3}, \quad y = b^2/a, \quad x = b^3 y^{-2} = a^2/b \text{ and } g\left(a^2/b, b^2/a\right) = 3ab.$$

The tangent plane is horizontal at $\left(a^2/b, b^2/a, 3ab\right)$.

21. Set $z = g(x, y) = xy$.   Then, $\nabla g = y\mathbf{i} + x\mathbf{j}$.

$$\nabla g = 0 \implies x = y = 0.$$

The tangent plane is horizontal at $(0, 0, 0)$.

23. Set $z = g(x, y) = 2x^2 + 2xy - y^2 - 5x + 3y - 2$.   Then,

$$\nabla g = (4x + 2y - 5)\mathbf{i} + (2x - 2y + 3)\mathbf{j}.$$

$$\nabla g = 0 \implies 4x + 2y - 5 = 0 = 2x - 2y + 3 \implies x = \tfrac{1}{3}, \quad y = \tfrac{11}{6}.$$

The tangent plane is horizontal at $\left(\frac{1}{3}, \frac{11}{6}, -\frac{1}{12}\right)$.

25. $\dfrac{x - x_0}{(\partial f/\partial x)(x_0, y_0, z_0)} = \dfrac{y - y_0}{(\partial f/\partial y)(x_0, y_0, z_0)} = \dfrac{z - z_0}{(\partial f/\partial z)(x_0, y_0, z_0)}$

27. Since the tangent planes meet at right angles, the normals ∇F and ∇G meet at right angles:

$$\frac{\partial F}{\partial x}\frac{\partial G}{\partial x} + \frac{\partial F}{\partial y}\frac{\partial G}{\partial y} + \frac{\partial F}{\partial z}\frac{\partial G}{\partial z} = 0.$$

29. The tangent plane at an arbitrary point (x_0, y_0, z_0) has equation

$$y_0 z_0 (x - x_0) + x_0 z_0 (y - y_0) + x_0 y_0 (z - z_0) = 0,$$

which simplifies to

$$y_0 z_0 x + x_0 z_0 y + x_0 y_0 z = 3x_0 y_0 z_0 \quad \text{and thus to} \quad \frac{x}{3x_0} + \frac{y}{3y_0} + \frac{z}{3z_0} = 1.$$

The volume of the pyramid is

$$V = \frac{1}{3}Bh = \frac{1}{3}\left[\frac{(3x_0)(3y_0)}{2}\right](3z_0) = \frac{9}{2}x_0 y_0 z_0 = \frac{9}{2}a^3.$$

31. The point $(2, 3, -2)$ is the tip of $\mathbf{r}(1)$.

Since $\quad \mathbf{r}'(t) = 2\mathbf{i} - \frac{3}{t^2}\mathbf{j} - 4t\mathbf{k}, \quad$ we have $\quad \mathbf{r}'(1) = 2\mathbf{i} - 3\mathbf{j} - 4\mathbf{k}.$

Now set $\quad f(x, y, z) = x^2 + y^2 + 3z^2 - 25.$ The function has gradient $2x\mathbf{i} + 2y\mathbf{j} + 6z\mathbf{k}.$

At the point $(2, 3, -2)$,

$$\nabla f = 2(2\mathbf{i} + 3\mathbf{j} - 6\mathbf{k}).$$

The angle θ between $\mathbf{r}'(1)$ and the gradient gives

$$\cos\theta = \frac{(2\mathbf{i} - 3\mathbf{j} - 4\mathbf{k})}{\sqrt{29}} \cdot \frac{(2\mathbf{i} + 3\mathbf{j} - 6\mathbf{k})}{7} = \frac{19}{7\sqrt{29}} \cong 0.504.$$

Therefore $\theta \cong 1.043$ radians. The angle between the curve and the plane is

$$\frac{\pi}{2} - \theta \cong 1.571 - 1.043 \cong 0.528 \text{ radians.}$$

33. Set $\quad f(x, y, z) = x^2 y^2 + 2x + z^3.$ Then,

$$\nabla f = (2xy^2 + 2)\mathbf{i} + 2x^2 y\mathbf{j} + 3z^2\mathbf{k}, \quad \nabla f(2, 1, 2) = 6\mathbf{i} + 8\mathbf{j} + 12\mathbf{k}.$$

The plane tangent to $f(x, y, z) = 16$ at $(2, 1, 2)$ has equation

$$6(x - 2) + 8(y - 1) + 12(z - 2) = 0, \quad 3x + 4y + 6z = 22.$$

Next, set $\quad g(x, y, z) = 3x^2 + y^2 - 2z.$ Then,

$$\nabla g = 6x\mathbf{i} + 2y\mathbf{j} - 2\mathbf{k}, \quad \nabla g(2, 1, 2) = 12\mathbf{i} + 2\mathbf{j} - 2\mathbf{k}.$$

The plane tangent to $g(x, y, z) = 9$ at $(2, 1, 2)$ is

$$12(x-2)+2(y-1)-2(z-2)=0, \quad 6x+y-z=11.$$

35.   The gradient to the sphere at $(1,1,2)$ is

$$2x\mathbf{i}+(2y-4)\mathbf{j}+(2z-2)\mathbf{k}=2\mathbf{i}-2\mathbf{j}+2\mathbf{k}.$$

The gradient to the paraboloid at $(1,1,2)$ is

$$6x\mathbf{i}+4y\mathbf{j}-2\mathbf{k}=6\mathbf{i}+4\mathbf{j}-2\mathbf{k}.$$

Since

$$(2\mathbf{i}-2\mathbf{j}+2\mathbf{k})\cdot(6\mathbf{i}+4\mathbf{j}-2\mathbf{k})=0,$$

the surfaces intersect at right angles.

37.   (a) $3x+4y+6=0$ since plane p is vertical.

(b) $y=-\frac{1}{4}(3x+6)=-\frac{1}{4}[3(4t-2)+6]=-3t$

$z=x^2+3y^2+2=(4t-2)^2+3(-3t)^2+2=43t^2-16t+6$

$\mathbf{r}(t)=(4t-2)\mathbf{i}-3t\mathbf{j}+(43t^2-16t+6)\mathbf{k}$

(c)   From part (b) the tip of $\mathbf{r}(1)$ is $(2,-3,33)$.   We take

$\mathbf{r}'(1)=4\mathbf{i}-3\mathbf{j}+70\mathbf{j}$ as $\mathbf{d}$ to write

$$\mathbf{R}(s)=(2\mathbf{i}-3\mathbf{j}+33\mathbf{k})+s(4\mathbf{i}-3\mathbf{j}+70\mathbf{k}).$$

(d)   Set $g(x,y)=x^2+3y^2+2$.   Then,

$$\nabla g=2x\mathbf{i}+6y\mathbf{j} \quad \text{and} \quad \nabla g(2,-3)=4\mathbf{i}-18\mathbf{j}.$$

An equation for the plane tangent to $z=g(x,y)$ at $(2,-3,33)$ is

$z-33=4(x-2)-18(y+3)$ which reduces to $4x-18y-z=29.$

(e)   Substituting t for x in the equations for p and p_1, we obtain

$$3t+4y+6=0 \quad \text{and} \quad 4t-18y-z=29.$$

From the first equation

$$y=-\tfrac{3}{4}(t+2)$$

and then from the second equation

$$z = 4t - 18\left[-\tfrac{3}{4}(t+2)\right] - 29 = \tfrac{35}{2}t - 2.$$

Thus,

$$(*) \qquad \mathbf{r}(t) = t\mathbf{i} - (\tfrac{3}{4}t + \tfrac{3}{2})\mathbf{j} + (\tfrac{35}{2}t - 2)\mathbf{k}.$$

Lines l and l' are the same. To see this, consider how l and l' are formed; to assure yourself, replace t in $(*)$ by $4s + 2$ to obtain $\mathbf{R}(s)$ found in part (c).

SECTION 15.6

1. $\nabla f(x,y) = (2-2x)\mathbf{i} - 2y\mathbf{j} = \mathbf{0}$ only at $(1,0)$.

    The difference
    $$f(1+h, k) - f(1,0) = \left[2(1+h) - (1+h)^2 - k^2\right] - 1 = -h^2 - k^2$$
    is negative for all small h and k; there is a local maximum of 1 at $(1,0)$.

3. $\nabla f(x,y) = (2-2x)\mathbf{i} + (2+2y)\mathbf{j} = \mathbf{0}$ only at $(1,-1)$.

    The difference
    $$f(1+h, -1+k) - f(1,-1)$$
    $$= [2(1+h) + 2(-1+k) - (1+h)^2 + (-1+k)^2 + 5] - 5 = -h^2 + k^2$$
    does not keep a constant sign for small h and k; $(1,-1)$ is a saddle point.

5. $\nabla f(x,y) = (2x+y+3)\mathbf{i} + (x+2y)\mathbf{j} = \mathbf{0}$ only at $(-2,1)$.

    The difference
    $$f(-2+h, 1+k) - f(-2,1)$$
    $$= [(-2+h)^2 + (-2+h)(1+k) + (1+k)^2 + 3(-2+h) + 1] - (-2) = h^2 + hk + k^2$$
    is positive for all small h and k. To see this, note that
    $$h^2 + hk + k^2 \geq h^2 + k^2 - |h|\,|k| > 0;$$
    there is a local minimum of -2 at $(-2,1)$.

7. $\nabla f = (3x^2 - 3)\mathbf{i} + \mathbf{j}$ is never $\mathbf{0}$; there are no stationary points and no local extreme values.

9. $\nabla f = (2x+y-3)\mathbf{i} + (x+2y-3)\mathbf{j} = \mathbf{0}$ only at $(1,1)$.

    The difference
    $$f(1+h, 1+k) - f(1,1) = [(1+h)^2 + (1+h)(1+k) + (1+k)^2 - 3(1+h) - 3(1+k)] - (-3) = h^2 + hk + k^2$$
    is positive for all small h and k. (See solution to Exercise 5 for details.) There is a local minimum of -3 at $(1,1)$.

11. $\nabla f = (2x + y + 2)\mathbf{i} + (x + 2)\mathbf{j} = 0$ only at $(-2, 2)$.

The difference

$$f(-2 + h,\, 2 + k) - f(-2, 2) = [(-2 + h)^2 + (-2 + h)(2 + k) + 2(-2 + h) + 2(2 + k) + 1] - 1$$

$$= h^2 + hk = h(h + k)$$

does not keep a constant sign for all small h and k. To see this, suppose $h > 0$. Then, $h(h + k)$ is positive when $k > -h$ and negative when $k < -h$. The point $(-2, 2)$ is a saddle point.

13. $\nabla f = (12x^2 y - 4y^3)\mathbf{i} + \left(4x^3 - 12xy^2\right)\mathbf{j} = 0$ only at $(0, 0)$.

The difference
$$f(h, k) - f(0, 0) = 4h^3 k - 4hk^3 = 4hk(h^2 - k^2)$$

does not keep a constant sign for all small h and k. To see this, suppose h and k are positive. Then, $4hk(h^2 - k^2)$ is positive when $h > k$ and negative when $h < k$. The origin is a saddle point.

15. $\nabla f = \dfrac{1}{\left(x^2 + y^2\right)^{3/2}}\,(-x\mathbf{i} - y\mathbf{j})$ is never 0 on D. Note that $f(x, y)$ is the reciprocal of the distance of (x, y) from the origin. The point of D closest to the origin (draw a figure) is $(1, 1)$. Therefore $f(1, 1) = 1/\sqrt{2}$ is the maximum value of f. The point of D furthest from the origin is $(3, 4)$. Therefore $f(3, 4) = 1/5$ is the least value taken on by f.

17. $\nabla f = 8x\mathbf{i} - 18y\mathbf{j} = 0$ only at $(0, 0)$.   Since the difference
$$f(h, k) - f(0, 0) = 4h^2 - 9k^2$$

does not keep a constant sign for all small h and k, the point $(0, 0)$ is a saddle point. It follows that f takes on no extreme values on the interior of D.

The boundary of D consists of the vertical lines $x = -1$ and $x = 1$. Note that

$$f(-1, y) = 4 - 9y^2 \text{ and } f(1, y) = 4 - 9y^2.$$

It is clear then that f takes on a maximum value of 4 [at $(-1, 0)$ and $(1, 0)$] but no minimum value.

19. $\nabla f = 2(x - 1)\mathbf{i} + 2(y - 1)\mathbf{j} = 0$ only at $(1, 1)$. As the sum of two squares, $f(x, y) \geq 0$. Thus, $f(1, 1) = 0$ is a minimum. To examine the behavior of f on the boundary of D, we note that f represents the square of the distance between (x, y) and $(1, 1)$. Thus, f is maximal at the point of the boundary furthest from $(1, 1)$. This is the point $\left(-\sqrt{2},\, -\sqrt{2}\right)$; the maximum value of f is $f\left(-\sqrt{2},\, -\sqrt{2}\right) = 6 + 4\sqrt{2}$.

21. $\nabla f = 2(x-y)\mathbf{i} - 2(x-y)\mathbf{j} = \mathbf{0}$ at each point of the line segment $y = x$ from $(0,0)$ to $(4,4)$. Since $f(x,x) = 0$ and $f(x,y) \geq 0$, f takes on its minimum of 0 at each of these points.

Next we consider the boundary of D.   We parametrize each side of the triangle:

$$C_1 : \mathbf{r}_1(t) = t\mathbf{j}, \quad t \in [0,12]$$
$$C_2 : \mathbf{r}_2(t) = t\mathbf{i}, \quad t \in [0,6]$$
$$C_3 : \mathbf{r}_3(t) = t\mathbf{i} + (12 - 2t)\mathbf{j}, \quad t \in [0,6]$$

and observe from

$$f(\mathbf{r}_1(t)) = t^2, \quad t \in [0,12]$$
$$f(\mathbf{r}_2(t)) = t^2, \quad t \in [0,6]$$
$$f(\mathbf{r}_3(t)) = (3t - 12)^2, \quad t \in [0,6]$$

that f takes on its maximum of 144 at the point $(0,12)$.

23. $\nabla f = 2(x-4)\mathbf{i} + 2y\mathbf{j}$ is never $\mathbf{0}$ at an interior point of D. Next we examine f on the boundary of D:

$$C_1 : \mathbf{r}_1(t) = t\mathbf{i} + 4t\mathbf{j}, \quad t \in [0,2,],$$
$$C_2 : \mathbf{r}_2(t) = t\mathbf{i} + t^3\mathbf{j}, \quad t \in [0,2].$$

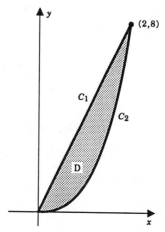

Note that

$$f_1(t) = f(\mathbf{r}_1(t)) = 17t^2 - 8t + 16,$$
$$f_2(t) = f(\mathbf{r}_2(t)) = (t - 4)^2 + t^6.$$

Next

$$f_1'(t) = 34t - 8 = 0 \quad \Longrightarrow \quad t = 4/17 \quad \text{and gives} \quad x = 4/17, \; y = 16/17$$

and

$$f_2'(t) = 6t^5 + 2t - 8 = 0 \quad \Longrightarrow \quad t = 1 \quad \text{and gives} \quad x = 1, \; y = 1.$$

The extreme values of f can be culled from the following list:

$$f(0,0) = 16, \quad f(2,8) = 68, \quad f\left(\tfrac{4}{17}, \tfrac{16}{17}\right) = \tfrac{256}{17}, \quad f(1,1) = 10.$$

We see that $f(1,1) = 10$ is the absolute minimum and $f(2,8)$ is the absolute maximum.

25. Let $P(x,y,z)$ be a point in the plane. We want to find the minimum of $f(x,y,z) = \sqrt{x^2 + y^2 + z^2}$.

However, it is sufficient to minimize the square of the distance: $F(x, y, z) = x^2 + y^2 + z^2$. It is clear that F has a minimum value, but no maximum value. Since P lies in the plane, $2x - y + 2z = 16$ which implies $y = 2x + 2z - 16 = 2(x + z - 8)$. Thus, we want to find the minimum value of

$$F(x, z) = x^2 + 4(x + z - 8)^2 + z^2$$

Now,

$$\nabla F = [2x + 8(x + z - 8)]\, wi + [8(x + z - 8)]\, \mathbf{k}$$

The gradient is $\mathbf{0}$ when

$$2x + 8(x + z - 8) = 0 \quad \text{and} \quad 8(x + z - 8) + 2z = 0$$

The only solution to this pair of equations is: $x = z = \dfrac{32}{9}$, from which it follows that $y = -\dfrac{16}{9}$.

The point in the plane that is closest to the origin is $P\left(\frac{32}{9}, -\frac{16}{9}, \frac{32}{9}\right)$.

The distance from the origin to the plane is: $F(P) = \frac{16}{3}$.

Check using (12.6.7): $d(P, 0) = \dfrac{|2 \cdot 0 - 0 + 2 \cdot 0 - 16|}{\sqrt{2^2 + (-1)^2 + 2^2}} = \dfrac{16}{3}$.

27. Using the hint, we want to find the maximum value of $f(x, y) = 18xy - x^2y - xy^2$.

The gradient of f is:

$$\nabla D = \left(18y - 2xy - y^2\right)\mathbf{i} + \left(18x - x^2 - 2xy\right)\mathbf{j}$$

The gradient is $\mathbf{0}$ when

$$18y - 2xy - y^2 = 0 \quad \text{and} \quad 18x - x^2 - 2xy = 0$$

The solution set of this pair of equations is: $(0,0)$, $(18,0)$, $(0,18)$, $(6,6)$.

It is easy to verify that f is a maximum when $x = y = 6$. The three numbers that satisfy $x+y+z = 18$ and maximize the product xyz are: $x = 6$, $y = 6$, $z = 6$.

29. Let the dimensions of the box be: length $= x$, width $= y$, height $= z$.

Then $V = xyz = 12 \implies z = \dfrac{12}{xy}$.

The surface area $S = xy + 2xz + 2yx = xy + \dfrac{24}{y} + \dfrac{24}{x}$. Thus, we want to minimize

$$S(x, y) = xy + \frac{24}{x} + \frac{24}{y}$$

The gradient of S is:

$$\left(y - \frac{24}{x^2}\right)\mathbf{i} + \left(x - \frac{24}{y^2}\right)\mathbf{j}$$

The gradient is $\mathbf{0}$ when

$$y - \frac{24}{x^2} = 0 \quad \text{and} \quad x - \frac{24}{y^2} = 0$$

Solving these equations simultaneously, we get $x = y = 2\sqrt[3]{3}$ which implies $z = \sqrt[3]{3}$.

The dimensions of the box with minimum surface area are: $2\sqrt[3]{3} \times 2\sqrt[3]{3} \times \sqrt[3]{3}$.

31. (a) $\nabla f = \frac{1}{2}x\,\mathbf{i} - \frac{2}{9}y\mathbf{j} = 0$ only at $(0,0)$.

(b) The difference

$$f(h,k) - f(0,0) = \tfrac{1}{4}h^2 - \tfrac{1}{9}k^2$$

does not keep a constant sign for all small h and k; $(0,0)$ is a saddle point. The function has no local extreme values.

(c) Being the difference of two squares, f can be maximized by maximizing $\frac{1}{4}x^2$ and minimizing $\frac{1}{9}y^2$; $(1,0)$ and $(-1,0)$ give absolute maximum value $\frac{1}{4}$. Similarly, $(0,1)$ and $(0,-1)$ give absolute minimum value $-\frac{1}{9}$.

33.
$$f(x,y) = \sum_{i=1}^{3}\left[(x-x_i)^2 + (y-y_i)^2\right]$$

$$\nabla f(x,y) = 2\left[(3x - x_1 - x_2 - x_3)\mathbf{i} + (3y - y_1 - y_2 - y_3)\mathbf{j}\right]$$

$$\nabla f = 0 \quad \text{only at} \quad \left(\frac{x_1 + x_2 + x_3}{3}, \frac{y_1 + y_2 + y_3}{3}\right) = (x_0, y_0).$$

The difference $\quad f(x_0 + h,\, y_0 + k) - f(x_0, y_0)$

$$= \sum_{i=1}^{3}\left[(x_0 + h - x_i)^2 + (y_0 + k - y_i)^2 - (x_0 - x_i)^2 - (y_0 - y_i)^2\right]$$

$$= \sum_{i=1}^{3}\left[2h(x_0 - x_i) + h^2 + 2k(y_0 - y_i) + k^2\right]$$

$$= 2h(3x_0 - x_1 - x_2 - x_3) + 2k(3y_0 - y_1 - y_2 - y_3) + h^2 + k^2$$

$$= h^2 + h^2$$

is positive for all small h and k. Thus, f has its absolute minimum at (x_0, y_0).

35. From the geometry of the situation we recognize that we want to find the distance between $\left(\frac{1}{2}, \frac{1}{2}, \frac{1}{2}\right)$ and the point on the sphere at the end of the radius through the point $\left(\frac{1}{2}, \frac{1}{2}, \frac{1}{2}\right)$. This means the distance we want is the radius diminished by the distance between the origin (center) and $\left(\frac{1}{2}, \frac{1}{2}, \frac{1}{2}\right)$:

$$1 - \sqrt{\left(\tfrac{1}{2}\right)^2 + \left(\tfrac{1}{2}\right)^2 + \left(\tfrac{1}{2}\right)^2} = 1 - \tfrac{1}{2}\sqrt{3}.$$

37. $P(x,y) = 1000[(x-80)(320-x) + (y-60)(140-y)], \quad 80 \le x \le 320, \quad 60 \le y \le 140.$

$\nabla P = 1000[(400 - 2x)\mathbf{i} + (200 - 2y)\mathbf{j}] = 0 \quad$ only at $\quad (200, 100).$

The difference

$$P(200 + h,\, 100 + k) - P(200, 100)$$

$$= 1000[(120 + h)(120 - h) + (40 + k)(40 - k) - (120)^2 - (40)^2]$$

$$= -1000 \left(h^2 + k^2 \right)$$

is negative for all small h and k. The profit is maximized by setting the selling prices at \$2 per razor, \$1 per dozen blades.

39. $(0,0)$ saddle point; $(1,1)$ local maximum   **41.** $(1,0)$ local minimum; $(-1,0)$ local maximum

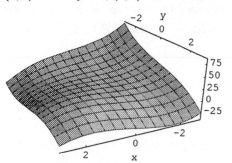   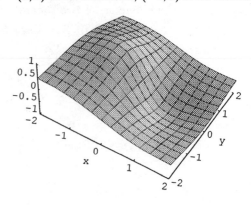

SECTION 15.7

1. $\nabla f = (2x + y - 6)\,\mathbf{i} + (x + 2y)\,\mathbf{j} = 0$ only at $(4, -2)$.

$f_{xx} = 2, \quad f_{xy} = 1, \quad f_{yy} = 2.$

At $(4, -2)$, $D = -3 < 0$ and $A = 2 > 0$ so we have a local min; the value is -10.

3. $\nabla f = (4 - 2x + y)\,\mathbf{i} + (2 + x - 2y)\,\mathbf{j} = 0$ only at $\left(\frac{10}{3}, \frac{8}{3} \right)$.

$f_{xx} = -2, \quad f_{xy} = 1, \quad f_{yy} = -2.$

At $\left(\frac{10}{3}, \frac{8}{3} \right)$, $D = -3 < 0$ and $A = -2 < 0$ so we have a local max; the value is $\frac{28}{3}$.

5. $\nabla f = (3x^2 - 6y)\mathbf{i} + \left(3y^2 - 6x \right)\mathbf{j} = 0$ at $(2,2)$ and $(0,0)$.

$f_{xx} = 6x, \quad f_{xy} = -6, \quad f_{yy} = 6y, \quad D = 36 - 36xy.$

At $(2,2)$, $D = -108 < 0$ and $A = 12 > 0$ so we have a local min; the value is -8.

At $(0,0)$, $D = 36 > 0$ so we have a saddle point.

7. $\nabla f = (3x^2 - 6y + 6)\mathbf{i} + (2y - 6x + 3)\mathbf{j} = 0$ at $\left(5, \frac{27}{2} \right)$ and $\left(1, \frac{3}{2} \right)$.

$f_{xx} = 6x, \quad f_{xy} = -6, \quad f_{yy} = 2, \quad D = 36 - 12x.$

At $\left(5, \frac{27}{2} \right)$, $D = -24 < 0$ and $A = 30 > 0$ so we have a local min; the value is $-\frac{117}{4}$.

At $\left(1, \frac{3}{2} \right)$, $D = 24 > 0$ so we have a saddle point.

9. $\nabla f = e^x \cos y\,\mathbf{i} - e^x \sin y\,\mathbf{j}$ is never $\mathbf{0}$; there are no stationary points and no local extreme values.

11. $\nabla f = \sin y\,\mathbf{i} + x \cos y\,\mathbf{j} = \mathbf{0}$ at $(0, n\pi)$ for all integral n.

$f_{xx} = 0, \quad f_{xy} = \cos y, \quad f_{yy} = -x \sin y, \quad D = \cos^2 y.$

Since $D = \cos^2 n\pi = 1 > 0$, each stationary point is a saddle point.

13. $\nabla f = (2xy + 1 + y^2)\mathbf{i} + (x^2 + 2xy + 1)\,\mathbf{j} = \mathbf{0}$ at $(1, -1)$ and $(-1, 1)$.

$f_{xx} = 2y, \quad f_{xy} = 2x + 2y, \quad f_{yy} = 2x, \quad D = 4(x + y)^2 - 4xy.$

At both $(1, -1)$ and $(-1, 1)$ we have saddle points since $D = 4 > 0$.

15. $\nabla f = (y - x^{-2})\mathbf{i} + (x - 8y^{-2})\,\mathbf{j} = \mathbf{0}$ only at $\left(\frac{1}{2}, 4\right)$.

$f_{xx} = 2x^{-3}, \quad f_{xy} = 1, \quad f_{yy} = 16y^{-3}, \quad D = 1 - 32x^{-3}y^{-3}.$

At $\left(\frac{1}{2}, 4\right)$, $D = -3 < 0$ and $A = 16 > 0$ so we have a local min; the value is 6.

17. $\nabla f = (y - x^{-2})\mathbf{i} + (x - y^{-2})\,\mathbf{j} = \mathbf{0}$ only at $(1, 1)$.

$f_{xx} = 2x^{-3}, \quad f_{xy} = 1, \quad f_{yy} = 2y^{-3}, \quad D = 1 - 4x^{-3}y^{-3}.$

At $(1, 1)$, $D = -3 < 0$ and $A = 2 > 0$ so we have a local min; the value is 3.

19. $\nabla f = \dfrac{2(x^2 - y^2 - 1)}{(x^2 + y^2 + 1)^2}\,\mathbf{i} + \dfrac{4xy}{(x^2 + y^2 + 1)^2}\,\mathbf{j} = \mathbf{0}$ at $(1, 0)$ and $(-1, 0)$.

$f_{xx} = \dfrac{-4x^3 + 12xy^2 + 12x}{(x^2 + y^2 + 1)^3}, \quad f_{xy} = \dfrac{4y^3 + 4y - 12x^2y}{(x^2 + y^2 + 1)^3}, \quad f_{yy} = \dfrac{4x^3 + 4xy^2 + 4x - 16xy^2}{(x^2 + y^2 + 1)^3}.$

At $(1, 0)$,

$$A = f_{xx}(1, 0) = 1 > 0, \quad B = f_{xy}(1, 0) = 0, \quad C = f_{yy}(1, 0) = 1, \quad D = -1 < 0.$$

Thus, $(1, 0)$ is a local min; $f(1, 0) = -1$.

At $(-1, 0)$,

$$A = f_{xx}(-1, 0) = -1 < 0, \quad B = f_{xy}(-1, 0) = 0, \quad C = f_{yy}(-1, 0) = -1, \quad D = -1 < 0.$$

Thus, $(1, 0)$ is a local max; $f(-1, 0) = 1$.

21. $\nabla f = (4x^3 - 4x)\,\mathbf{i} + 2y\,\mathbf{j} = \mathbf{0}$ at $(0, 0), (1, 0)$, and $(-1, 0)$.

$f_{xx} = 12x^2 - 4, \quad f_{xy} = 0, \quad f_{yy} = 2, \quad D = 8 - 24x^2.$

At $(0, 0)$, $D = 8 > 0$. Thus, $(0, 0)$ is a saddle point.

At $(\pm 1, 0)$, $D = -16 < 0$ and $A = 8 > 0$. Thus, the points $(1, 0)$ and $(-1, 0)$ are local minima; $f(\pm 1, 0) = -3$.

23. $\nabla f = [-\sin x + \sin(x + y)]\mathbf{i} + [-\sin y + \sin(x + y)]\mathbf{j} = \mathbf{0}$ at $(0, 0), (\pi/3, \pi/3)$ and $(-\pi/3, -\pi/3)$.

$$f_{xx} = -\cos x + \cos(x+y), \quad f_{xy} = \cos(x+y), \quad f_{yy} = -\cos y + \cos(x+y).$$

At $(0,0)$, $A = f_{xx}(0,0) = 0$, $B = f_{xy}(0,0) = 1$, $C = f_{yy}(0,0) = 0$, $D = 1 > 0$.

Thus $(0,0)$ is a saddle point.

At $(\pi/3, \pi/3)$, $A = f_{xx}(\pi/3, \pi/3) = -1 < 0$, $B = f_{xy}(\pi/3, \pi/3) = -\dfrac{1}{2}$,

$C = f_{yy}(\pi/3, \pi/3) = -1$, $D = -\dfrac{3}{4} < 0$,

so we have a local max; $f(\pi/3, \pi/3) = \dfrac{3}{2}$. By symmetry, $(-\pi/3, -\pi/3)$ is also a local max

and $f(-\pi/3, -\pi/3) = \dfrac{3}{2}$.

25. (a) $\nabla f = (2x + ky)\mathbf{i} + (2y + kx)\mathbf{j}$ and $\nabla f(0,0) = \mathbf{0}$ independent of the value of k.

  (b) $f_{xx} = 2$, $f_{xy} = k$, $f_{yy} = 2$, $D = k^2 - 4$. Thus, $D > 0$ for $|k| > 2$ and $(0,0)$ is a saddle point

  (c) $D = k^2 - 4 < 0$ for $|k| < 2$. Since $A = f_{xx} = 2 > 0$, $(0,0)$ is a local minimum.

  (d) The test is inconclusive when $D = k^2 - 4 = 0$ i.e., for $k = \pm 2$.

27. It is sufficient to minimize the square of the distance from a point $P(x, y, z)$ in the plane to $(-1, 1, 2)$.

  Set

$$f(x, y) = (x+1)^2 + (y-1)^2 + (z-2)^2$$
$$= (x+1)^2 + (y-1)^2 + \frac{1}{36}(2 - 2x + 3y)^2 \quad \left[\text{since } z = \frac{1}{6}(14 - 2x + 3y)\right]$$

$$\nabla f = \left[2(x+1) - \frac{1}{9}(2 - 2x + 3y)\right]\mathbf{i} + \left[2(y-1) + \frac{1}{6}(2 - 2x + 3y)\right]\mathbf{j} = \mathbf{0} \implies x = -\frac{5}{7}, \ y = \frac{4}{7}.$$

$$f_{xx} = \frac{20}{9} > 0, \quad f_{xy} = -\frac{1}{3}, \quad f_{yy} = \frac{5}{2}, \quad D = \frac{1}{9} - \frac{50}{9} = -\frac{49}{9} < 0.$$

  Thus, f has a local minimum at the point $(-5/7, 4/7)$. Now, $z = 20/7$ when $x = -5/7$ and $y = 4/7$.

  Thus, the point in the plane $2x - 3y + 6z = 14$ that is closest to $(-1, 1, 2)$ is $(-5/7, 4/7, 20/7)$.

29. $f(x, y) = xy(1 - x - y)$, $0 \le x \le 1$, $0 \le y \le 1 - x$.

  [dom (f) is the triangle with vertices $(0,0)$, $(1,0)$, $(0,1)$.]

  $\nabla f = (y - 2xy - y^2)\mathbf{i} + (x - 2xy - x^2)\mathbf{j} = \mathbf{0} \implies x = y = \frac{1}{3}$.

  (Note that $[0,0]$ is not an interior point of the domain of f.)

  $f_{xx} = -2x$, $f_{xy} = 1 - 2x - 2y$, $f_{yy} = -2x$, $D = (1 - 2x - 2y)^2 - 4xy$.

  At $\left(\frac{1}{3}, \frac{1}{3}\right)$, $D = -\frac{1}{3} < 0$ and $A > 0$ so we have a local max; the value is $1/27$.

  Since $f(x, y) = 0$ at each point on the boundary of the domain, the local max of $1/27$ is also the absolute max.

31. $f(x, y) = (x-1)^2 + (y-2)^2 + z^2 = (x-1)^2 + (y-2)^2 + x^2 + 2y^2 \quad \left[\text{since } z = \sqrt{x^2 + 2y^2}\right]$

  $\nabla f = [2(x-1) + 2x]\mathbf{i} + [2(y-2) + 4y]\mathbf{j} = \mathbf{0} \implies x = \frac{1}{2}, \ y = \frac{2}{3}$.

$f_{xx} = 4 > 0,$ $f_{xy} = 0,$ $f_{yy} = 6,$ $D = -24 < 0.$ Thus, f has a local minimum at $(1/2, 2/3)$.

The shortest distance from $(1, 2, 0)$ to the cone is $f\left(\frac{1}{2}, \frac{2}{3}\right) = \frac{1}{6}\sqrt{114}$

33. (a) $f(x, y) = 0$ along the plane curve $y = x^{2/3}$.
Since $f(x, y)$ is positive for points below the
curve and is negative for points above the
curve, there is a saddle point at the origin.

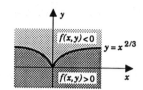

(b) $\nabla f = (ye^{xy} - 2\sin(x + y))\mathbf{i} + (xe^{xy} - 2\sin(x + y))\mathbf{j}$

$$f_{xx} = -2\cos(x + y) + y^2 e^{xy},$$

$$f_{xy} = -2\cos(x + y) + e^{xy}(1 + xy),$$

$$f_{yy} = -2\cos(x + y) + x^2 e^{xy}.$$

At the origin $A = -2$, $B = -1$, $C = -2$, and $D = -3$. We have a local max; the value is 3.

35.

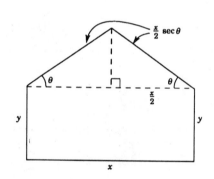

$$A = xy + \frac{1}{2}x\left(\frac{x}{2}\tan\theta\right),$$

$$P = x + 2y + 2\left(\frac{x}{2}\sec\theta\right),$$

$$0 < \theta < \frac{1}{2}\pi, \quad 0 < x < \frac{P}{1 + \sec\theta}.$$

$$A(x, \theta) = \frac{1}{2}x(P - x - x\sec\theta) + \frac{1}{4}x^2\tan\theta,$$

$$\nabla A = \left(\frac{P}{2} - x - x\sec\theta + \frac{x}{2}\tan\theta\right)\mathbf{i} + \left(\frac{x^2}{4}\sec^2\theta - \frac{x^2}{2}\sec\theta\tan\theta\right)\mathbf{j},$$

$$\nabla A = \frac{1}{2}[P + x(\tan\theta - 2\sec\theta - 2)]\mathbf{i} + \frac{x^2}{4}\sec\theta\,(\sec\theta - 2\tan\theta)\mathbf{j}.$$

From $\dfrac{\partial A}{\partial \theta} = 0$ we get $\theta = \frac{1}{6}\pi$ and then from $\dfrac{\partial A}{\partial x} = 0$ we get

$$P + x\left(\tfrac{1}{3}\sqrt{3} - \tfrac{4}{3}\sqrt{3} - 2\right) = 0 \quad \text{so that} \quad x = (2 - \sqrt{3})P.$$

Next,

$$A_{xx} = \tfrac{1}{2}(\tan\theta - 2\sec\theta - 2),$$

$$A_{x\theta} = \frac{x}{2}\sec\theta\,(\sec\theta - 2\tan\theta),$$

$$A_{\theta\theta} = \frac{x^2}{2}\sec\theta\,(\sec\theta\tan\theta - \sec^2\theta - \tan^2\theta).$$

By the second-partials test

$$A = -\tfrac{1}{2}(2+\sqrt{3}), \quad B = 0, \quad C = -\tfrac{1}{3}P^2\sqrt{3}(2-\sqrt{3})^2, \quad D < 0.$$

The area is a maximum when $\theta = \tfrac{1}{6}\pi$, $x = (2-\sqrt{3})P$ and $y = \tfrac{1}{6}(3-\sqrt{3})P$.

37. From $\qquad x = \tfrac{1}{2}y = \tfrac{1}{3}z = t \quad$ and $\quad x = y - 2 = z = s$

we take $\qquad\qquad\qquad (t, 2t, 3t) \quad$ and $\quad (s, 2+s, s)$

as arbitrary points on the lines. It suffices to minimize the square of the distance between these points:

$$f(t,s) = (t-s)^2 + (2t-2-s)^2 + (3t-s)^2$$

$$= 14t^2 - 12ts + 3s^2 - 8t + 4s + 4, \qquad t, s \text{ real.}$$

$$\nabla f = (28t - 12s - 8)\mathbf{i} + (-12t + 6s + 4)\mathbf{j}; \qquad \nabla f = 0 \implies t = 0, \ s = -2/3.$$

$$f_{tt} = 28, \quad f_{ts} = -12, \quad f_{ss} = 6, \quad D = (-12)^2 - 6(28) = -24 < 0.$$

By the second-partials test, the distance is a minimum when $t = 0$, $s = -2/3$; the nature of the problem tells us the minimum is absolute. The distance is $\sqrt{f(0, 2/3)} = \tfrac{2}{3}\sqrt{6}$.

39.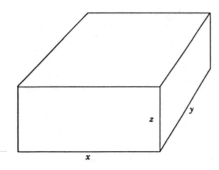

$$96 = xyz,$$

$$C = 30xy + 10(2xz + 2yz)$$

$$= 30xy + 20(x+y)\frac{96}{xy}.$$

$$C(x,y) = 30\left[xy + \frac{64}{x} + \frac{64}{y}\right],$$

$$\nabla C = 30(y - 64x^{-2})\mathbf{i} + 30(x - 64y^{-2})\mathbf{j} = 0 \implies x = y = 4.$$

$$C_{xx} = 128x^{-3}, \quad C_{xy} = 1, \quad C_{yy} = 128y^{-3}.$$

When $x = y = 4$, we have $D = -3 < 0$ and $A = 2 > 0$ so the cost is minimized by making the dimensions of the crate $4 \times 4 \times 6$ meters.

41. Let x, y and z be the length, width and height of the box. The surface area is given by

$$S = 2xy + 2xz + 2yz, \quad \text{so} \quad z = \frac{S - 2xy}{2(x+y)}, \quad \text{where } S \text{ is a constant, and } x, y, z > 0.$$

Now, the volume $V = xyz$ is given by:

$$V(x, y) = xy \left[\frac{S - 2xy}{2(x + y)} \right]$$

and

$$\nabla V = y \left\{ \left[\frac{S - 2xy}{2(x + y)} \right] + xy \, \frac{2(x + y)(-2y) - (S - 2xy)(2)}{4(x + y)^2} \right\} \mathbf{i}$$

$$+ \left\{ x \left[\frac{S - 2xy}{2(x + y)} \right] + xy \, \frac{2(x + y)(-2x) - (S - 2xy)(2)}{4(x + y)^2} \right\} \mathbf{j}$$

Setting $\dfrac{\partial V}{\partial x} = \dfrac{\partial V}{\partial y} = 0$ and simplifying, we get the pair of equations

$$2S - 4x^2 - 8xy = 0$$

$$2S - 4y^2 - 8xy = 0$$

from which it follows that $x = y = \sqrt{S/6}$. From practical considerations, we conclude that V has a maximum value at $(\sqrt{S/6}, \sqrt{S/6})$. Substituting these values into the equation for z, we get $z = \sqrt{S/6}$ and so the box of maximum volume is a cube.

43. (a) $f(m, b) = [2 - b]^2 + [-5 - (m + b)]^2 + [4 - (2m + b)]^2$.

$f_m = 10m + 6b - 6$, $\quad f_b = 6m + 6b - 2$; $\quad f_m = f_b = 0 \implies m = 1$, $b = -\frac{2}{3}$.

$f_{mm} = 10$, $\quad f_{mb} = 6$, $\quad f_{bb} = 6$, $\quad D = -24 < 0 \implies$ a min.

Answer: the line $y = x - \frac{2}{3}$.

(b) $f(\alpha, \beta) = [2 - \beta]^2 + [-5 - (\alpha + \beta)]^2 + [4 - (4\alpha + \beta)]^2$.

$f_\alpha = 34\alpha + 10\beta - 22$, $\quad f_\beta = 10\alpha + 6\beta - 2$; $\quad f_\alpha = f_\beta = 0 \implies \left[\begin{aligned} \alpha &= \tfrac{14}{13} \\ \beta &= -\tfrac{19}{13} \end{aligned} \right]$.

$f_{\alpha\alpha} = 34$, $\quad f_{\alpha\beta} = 10$, $\quad f_{\beta\beta} = 6$, $\quad D = -104 < 0 \implies$ a min.

Answer: the parabola $y = \frac{1}{13}\left(14x^2 - 19\right)$.

45. (a) Let x and y be the cross-sectional measurements of the box, and let l be its length. Then

$$V = xyl, \quad \text{where} \quad 2x + 2y + l \le 108, \quad x, y > 0$$

To maximize V we will obviously take $2x + 2y + l = 108$. Therefore, $V(x, y) = xy(108 - 2x - 2y)$ and

$$\nabla V = [y(108 - 2x - 2y) - 2xy]\,\mathbf{i} + [x(108 - 2x - 2y) - 2xy]\,\mathbf{j}$$

Setting $\dfrac{\partial V}{\partial x} = \dfrac{\partial V}{\partial y} = 0$, we get the pair of equations

$$\frac{\partial V}{\partial x} = 108y - 4xy - 2y^2 = 0$$

$$\frac{\partial V}{\partial y} = 108x - 4xy - 2x^2 = 0$$

from which it follows that $x = y = 18 \implies l = 36$.

Now, at $(18, 18)$, we have

$$A = V_{xx} = -4y = -72 < 0, \quad B = V_{xy} = 108 - 4x - 4y = -36,$$

$$C = V_{yy} = -4x = -72, \quad \text{and} \quad D = (36)^2 - (72)^2 < 0.$$

Thus, V is a maximum when $x = y = 18$ and $l = 36$.

(b)   Let r be the radius of the tube and let l be its length.

Then

$$V = \pi r^2 l, \quad \text{where} \quad 2\pi r + l \le 108, \quad r > 0$$

To maximize V we take $2\pi r + l = 108$. Then $V(r) = \pi r^2 (108 - 2\pi r) = 108\pi r^2 - 2\pi^2 r^3$. Now

$$\frac{dV}{dr} = 216\pi r - 6\pi^2 r^2$$

Setting $\dfrac{dV}{dr} = 0$, we get

$$216\pi r - 6\pi^2 r^2 = 0 \quad \Longrightarrow \quad r = \frac{36}{\pi} \quad \Longrightarrow \quad l = 36$$

Now, at $r = 36/\pi$, we have

$$\frac{d^2 V}{dr^2} = 216\pi - 12\pi^2 \frac{36}{\pi} = -216\pi < 0$$

Thus, V is a maximum when $r = 36/\pi$ and $l = 36$.

47.   Let S denote the cross-sectional area. Then

$$S = \frac{1}{2}(12 - 2x + 12 - 2x + 2x \cos\theta)\, x \sin\theta = 12x \sin\theta - 2x^2 \sin\theta + \frac{1}{2}x^2 \sin 2\theta,$$

where $0 < x < 6, \quad 0 < \theta < \pi/2$

Now,

$$\nabla S = (12 \sin\theta - 4x \sin\theta + x \sin 2\theta)\mathbf{i} + (12x \cos\theta - 2x^2 \cos\theta + x^2 \cos 2\theta)\mathbf{j}$$

Setting $\dfrac{\partial S}{\partial x} = \dfrac{\partial S}{\partial \theta} = 0$, we get the pair of equations

$$12 \sin\theta - 4x \sin\theta + x \sin 2\theta = 0$$

$$12x \cos\theta - 2x^2 \cos\theta + x^2 \cos 2\theta = 0$$

from which it follows that $x = 4, \theta = \pi/3$.

Now, at $(4, \pi/3)$, we have

$$A = S_{xx} = -4 \sin\theta + \sin 2\theta = -\frac{3}{2}\sqrt{3}, \quad B = S_{x\theta} = 12 \cos\theta - 4x \cos\theta + 2x \cos 2\theta = -6,$$

$$C = S_{\theta\theta} = -12x \sin\theta + 2x^2 \sin\theta - 2x^2 \sin 2\theta = -24\sqrt{3} \quad \text{and} \quad D = 36 - 108 < 0.$$

Thus, S is a maximum when $x = 4$ and $\theta = \pi/3$.

SECTION 15.8

1.
$$f(x,y) = x^2 + y^2, \qquad g(x,y) = xy - 1$$

$$\nabla f = 2x\mathbf{i} + 2y\mathbf{j}, \qquad \nabla g = y\mathbf{i} + x\mathbf{j}.$$

$$\nabla f = \lambda \nabla g \implies 2x = \lambda y \text{ and } 2y = \lambda x.$$

Multiplying the first equation by x and the second equation by y, we get

$$2x^2 = \lambda xy = 2y^2.$$

Thus, $x = \pm y$. From $g(x,y) = 0$ we conclude that $x = y = \pm 1$. The points $(1,1)$ and $(-1,-1)$ clearly give a minimum, since f represents the square of the distance of a point on the hyperbola from the origin. The minimum is 2.

3.
$$f(x,y) = xy, \qquad g(x,y) = b^2 x^2 + a^2 y^2 - a^2 b^2$$

$$\nabla f = y\mathbf{i} + x\mathbf{j}, \qquad \nabla g = 2b^2 x\mathbf{i} + 2a^2 y\mathbf{j}.$$

$$\nabla f = \lambda \nabla g \implies y = 2\lambda b^2 x \text{ and } x = 2\lambda a^2 y.$$

Multiplying the first equation by $a^2 y$ and the second equation by $b^2 x$, we get

$$a^2 y^2 = 2\lambda a^2 b^2 xy = b^2 x^2.$$

Thus, $ay = \pm bx$. From $g(x,y) = 0$ we conclude that $x = \pm \frac{1}{2} a\sqrt{2}$ and $y = \pm \frac{1}{2} b\sqrt{2}$.

Since f is continuous and the ellipse is closed and bounded, the minimum exists. It occurs at $\left(\frac{1}{2}a\sqrt{2}, -\frac{1}{2}b\sqrt{2}\right)$ and $\left(-\frac{1}{2}a\sqrt{2}, \frac{1}{2}b\sqrt{2}\right)$; the minimum is $-\frac{1}{2}ab$.

5.   Since f is continuous and the ellipse is closed and bounded, the maximum exists.

$$f(x,y) = xy^2, \qquad g(x,y) = b^2 x^2 + a^2 y^2 - a^2 b^2$$

$$\nabla f = y^2\mathbf{i} + 2xy\mathbf{j}, \qquad \nabla g = 2b^2 x\mathbf{i} + 2a^2 y\mathbf{j}.$$

$$\nabla f = \lambda \nabla g \implies y^2 = 2\lambda b^2 x \text{ and } 2xy = 2\lambda a^2 y.$$

Multiplying the first equation by $a^2 y$ and the second equation by $b^2 x$, we get

$$a^2 y^3 = 2\lambda a^2 b^2 xy = 2b^2 x^2 y.$$

We can exclude $y = 0$; it clearly cannot produce the maximum. Thus,

$$a^2 y^2 = 2b^2 x^2 \text{ and, from } g(x,y) = 0, \ 3b^2 x^2 = a^2 b^2.$$

This gives us $x = \pm \frac{1}{3}\sqrt{3}\,a$ and $y = \pm \frac{1}{3}\sqrt{6}\,b$. This maximum occurs at $x = \frac{1}{3}\sqrt{3}\,a$, $y = \pm \frac{1}{3}\sqrt{6}\,b$; the value there is $\frac{2}{9}\sqrt{3}\,ab^2$.

7.   The given curve is closed and bounded. Since $x^2 + y^2$ represents the square of the distance from points on this curve to the origin, the maximum exists.

$$f(x,y) = x^2 + y^2, \qquad\qquad g(x,y) = x^4 + 7x^2y^2 + y^4 - 1$$

$$\nabla f = 2x\mathbf{i} + 2y\mathbf{j}, \qquad\qquad \nabla g = \left(4x^3 + 14xy^2\right)\mathbf{i} + \left(4y^3 + 14x^2y\right)\mathbf{j}.$$

We use the cross-product equation (16.8.4):

$$2x(4y^3 + 14x^2y) - 2y(4x^3 + 14xy^2) = 0,$$

$$20x^3y - 20xy^3 = 0,$$

$$xy(x^2 - y^2) = 0.$$

Thus, $x = 0$, $y = 0$, or $x = \pm y$. From $g(x, y) = 0$ we conclude that the points to examine are

$$(0, \pm 1), \quad (\pm 1, 0), \quad \left(\pm\tfrac{1}{3}\sqrt{3}, \pm\tfrac{1}{3}\sqrt{3}\right).$$

The value of f at each of the first four points is 1; the value at the last four points is 2/3. The maximum is 1.

9. The maximum exists since xyz is continuous and the ellipsoid is closed and bounded.

$$f(x, y, z) = xyz, \qquad\qquad g(x, y, z) = \frac{x^2}{a^2} + \frac{y^2}{b^2} + \frac{z^2}{c^2} - 1$$

$$\nabla f = yz\mathbf{i} + xz\mathbf{j} + xy\mathbf{k}, \qquad\qquad \nabla g = \frac{2x}{a^2}\mathbf{i} + \frac{2y}{b^2}\mathbf{j} + \frac{2z}{c^2}\mathbf{k}.$$

$$\nabla f = \lambda \nabla g \implies yz = \frac{2x}{a^2}\lambda, \quad xz = \frac{2y}{b^2}\lambda, \quad xy = \frac{2z}{c^2}\lambda.$$

We can assume x, y, z are non-zero, for otherwise $f(x, y, z) = 0$, which is clearly not a maximum. Then from the first two equations

$$\frac{yza^2}{x} = 2\lambda = \frac{xzb^2}{y} \quad \text{so that} \quad a^2y^2 = b^2x^2 \quad \text{or} \quad x^2 = \frac{a^2y^2}{b^2}.$$

Similarly from the second and third equations we get

$$b^2z^2 = c^2y^2 \quad \text{or} \quad z^2 = \frac{c^2y^2}{b^2}.$$

Substituting these expressions for x^2 and z^2 in $g(x, y, z) = 0$, we obtain

$$\frac{1}{a^2}\left[\frac{a^2y^2}{b^2}\right] + \frac{y^2}{b^2} + \frac{1}{c^2}\left[\frac{c^2y^2}{b^2}\right] - 1 = 0, \quad \frac{3y^2}{b^2} = 1, \quad y = \pm\frac{1}{3}b\sqrt{3}.$$

Then, $x = \pm\frac{1}{3}a\sqrt{3}$ and $z = \pm\frac{1}{3}c\sqrt{3}$. The maximum value is $\frac{1}{9}\sqrt{3}\,abc$.

11. Since the sphere is closed and bounded and $2x + 3y + 5z$ is continuous, the maximum exists.

$$f(x,y,z) = 2x + 3y + 5z, \qquad g(x,y,z) = x^2 + y^2 + z^2 - 19$$

$$\nabla f = 2\mathbf{i} + 3\mathbf{j} + 5\mathbf{k}, \qquad \nabla g = 2x\mathbf{i} + 2y\mathbf{j} + 2z\mathbf{k}.$$

$$\nabla f = \lambda \nabla g \implies 2 = 2\lambda x, \quad 3 = 2\lambda y, \quad 5 = 2\lambda z.$$

Since $\lambda \neq 0$ here, we solve the equations for x, y and z:

$$x = \frac{1}{\lambda}, \quad y = \frac{3}{2\lambda}, \quad z = \frac{5}{2\lambda},$$

and substitute these results in $g(x,y,z) = 0$ to obtain

$$\frac{1}{\lambda^2} + \frac{9}{4\lambda^2} + \frac{25}{4\lambda^2} - 19 = 0, \quad \frac{38}{4\lambda^2} - 19 = 0, \quad \lambda = \pm\frac{1}{2}\sqrt{2}.$$

The positive value of λ will produce positive values for x, y, z and thus the maximum for f. We get $x = \sqrt{2}, y = \frac{3}{2}\sqrt{2}, z = \frac{5}{2}\sqrt{2}$, and $2x + 3y + 5z = 19\sqrt{2}$.

13.
$$f(x,y,z) = xyz, \qquad g(x,y,z) = \frac{x}{a} + \frac{y}{b} + \frac{z}{c} - 1$$

$$\nabla f = yz\mathbf{i} + xz\mathbf{j} + xy\mathbf{k}, \qquad \nabla g = \frac{1}{a}\mathbf{i} + \frac{1}{b}\mathbf{j} + \frac{1}{c}\mathbf{k}.$$

$$\nabla f = \lambda \nabla g \implies yz = \frac{\lambda}{a}, \quad xz = \frac{\lambda}{b}, \quad xy = \frac{\lambda}{c}.$$

Multiplying these equations by x, y, z respectively, we obtain

$$xyz = \frac{\lambda x}{a}, \quad xyz = \frac{\lambda y}{b}, \quad xyz = \frac{\lambda z}{c}.$$

Adding these equations and using the fact that $g(x,y,z) = 0$, we have

$$3xyz = \lambda\left(\frac{x}{a} + \frac{y}{b} + \frac{z}{c}\right) = \lambda.$$

Since x, y, z are non-zero,

$$yz = \frac{\lambda}{a} = \frac{3xyz}{a}, \quad 1 = \frac{3x}{a}, \quad x = \frac{a}{3}.$$

Similarly, $y = \frac{b}{3}$ and $z = \frac{c}{3}$. The maximum is $\frac{1}{27}abc$.

15. It suffices to minimize the square of the distance from $(0,1)$ to the parabola. Clearly, the minimum exists.

$$f(x,y) = x^2 + (y-1)^2, \qquad g(x,y) = x^2 - 4y$$

$$\nabla f = 2x\mathbf{i} + 2(y-1)\mathbf{j}, \qquad \nabla g = 2x\mathbf{i} - 4\mathbf{j}.$$

We use the cross-product equation (15.8.4):

$$2x(-4) - 2x(2y - 2) = 0, \quad 4x + 4xy = 0, \quad x(y + 1) = 0.$$

Since $y \geq 0$, we have $x = 0$ and thus $y = 0$. The minimum is 1.

17. It suffices to maximize and minimize the square of the distance from $(2, 1, 2)$ to the sphere. Clearly, these extreme values exist.

$$f(x, y, z) = (x - 2)^2 + (y - 1)^2 + (z - 2)^2 \qquad\qquad g(x, y, z) = x^2 + y^2 + z^2 - 1$$

$$\nabla f = 2(x - 2)\mathbf{i} + 2(y - 1)\mathbf{j} + 2(z - 2)\mathbf{k}, \qquad\qquad \nabla g = 2x\,\mathbf{i} + 2y\,\mathbf{j} + 2z\,\mathbf{k}.$$

$$\nabla f = \lambda = \nabla g \implies 2(x - 2) = 2x\lambda, \;\; 2(y - 1) = 2y\lambda, \;\; 2(z - 2) = 2z\lambda$$

Thus,

$$x = \frac{2}{1 - \lambda}, \quad y = \frac{1}{1 - \lambda}, \quad z = \frac{2}{1 - \lambda}.$$

Using the fact that $x^2 + y^2 + z^2 = 1$, we have

$$\left(\frac{2}{1 - \lambda}\right)^2 + \left(\frac{1}{1 - \lambda}\right)^2 + \left(\frac{2}{1 - \lambda}\right)^2 = 1 \implies \lambda = -2, 4$$

At $\lambda = -2$, $(x, y, z) = (2/3, 1/3, 2/3)$ and $f(2/3, 1/3/2/3) = 4$

At $\lambda = 4$, $(x, y, z) = (-2/3, -1/3, -2/3)$ and $f(-2/3, -1/3/ - 2/3) = 16$

Thus, $(2/3, 1/3, 2/3)$ is the closest point and $(-2/3, -1/3, -2/3)$ is the furthest point.

19. $$f(x, y, z) = 3x - 2y + z, \qquad\qquad g(x, y, z) = x^2 + y^2 + z^2 - 14$$

$$\nabla f = 3\mathbf{i} - 2\mathbf{j} + \mathbf{k}, \qquad\qquad \nabla g = 2x\,\mathbf{i} + 2y\,\mathbf{j} + 2z\,\mathbf{k}.$$

$$\nabla f = \lambda \nabla g \implies 3 = 2x\lambda, \;\; -2 = 2y\lambda, \;\; 1 = 2z\lambda.$$

Thus,

$$x = \frac{3}{2\lambda}, \quad y = -\frac{1}{\lambda}, \quad z = \frac{1}{2\lambda}.$$

Using the fact that $x^2 + y^2 + z^2 = 14$, we have

$$\left(\frac{3}{2\lambda}\right)^2 + \left(-\frac{1}{\lambda}\right)^2 + \left(\frac{1}{2\lambda}\right)^2 = 14 \implies \lambda = \pm\frac{1}{2}.$$

At $\lambda = -\dfrac{1}{2}$, $(x, y, z) = (3, -2, 1)$ and $f(3, -2, 1) = 14$

At $\lambda = \dfrac{1}{2}$, $(x, y, z) = (-3, 2, -1)$ and $f(-3, 2, -1) = -14$

Thus, the maximum value of f on the sphere is 14.

21. It's easier to work with the square of the distance; the minimum certainly exists.

$$f(x, y, z) = x^2 + y^2 + z^2, \qquad\qquad g(x, y, z) = Ax + By + Cz + D$$

$$\nabla f = 2x\mathbf{i} + 2y\mathbf{j} + 2z\mathbf{k}, \qquad\qquad \nabla g = A\mathbf{i} + B\mathbf{j} + C\mathbf{k}.$$

$$\nabla f = \lambda \nabla g \quad \Longrightarrow \quad 2x = A\lambda, \quad 2y = B\lambda, \quad 2z = C\lambda.$$

Substituting these equations in $g(x,y,z) = 0$, we have

$$\frac{1}{2}\lambda\left(A^2 + B^2 + C^2\right) + D = 0, \quad \lambda \frac{-2D}{A^2 + B^2 + C^2}.$$

Thus, in turn,

$$x = \frac{-DA}{A^2 + B^2 + C^2}, \quad y = \frac{-DB}{A^2 + B^2 + C^2}, \quad z = \frac{-DC}{A^2 + B^2 + C^2}$$

so the minimum value of $\sqrt{x^2 + y^2 + z^2}$ is $|D|\left(A^2 + B^2 + C^2\right)^{-1/2}$.

23.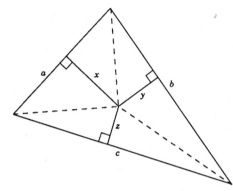

area $A = \frac{1}{2}ax + \frac{1}{2}by + \frac{1}{2}cz$.

The geometry suggests that

$$x^2 + y^2 + z^2$$

has a minimum.

$$f(x,y,z) = x^2 + y^2 + z^2, \qquad g(x,y,z) = ax + by + cz - 2A$$

$$\nabla f = 2x\mathbf{i} + 2y\mathbf{j} + 2z\mathbf{k}, \qquad \nabla g = a\mathbf{i} + b\mathbf{j} + c\mathbf{k}.$$

$$\nabla f = \lambda \nabla g \quad \Longrightarrow \quad 2x = a\lambda, \quad 2y = b\lambda, \quad 2z = c\lambda.$$

Solving these equations for x, y, z and substituting the results in $g(x,y,z) = 0$, we have

$$\frac{a^2\lambda}{2} + \frac{b^2\lambda}{2} + \frac{c^2\lambda}{2} - 2A = 0, \quad \lambda = \frac{4A}{a^2 + b^2 + c^2}$$

and thus

$$x = \frac{2aA}{a^2 + b^2 + c^2}, \quad y = \frac{2bA}{a^2 + b^2 + c^2}, \quad z = \frac{2cA}{a^2 + b^2 + c^2}.$$

The minimum is $4A^2(a^2 + b^2 + c^2)^{-1}$.

25. Since the curve is asymptotic to the line $y = x$ as $x \to -\infty$ and as $x \to \infty$, the maximum exists. The distance between the point (x,y) and the line $y - x = 0$ is given by

$$\frac{|y - x|}{\sqrt{1 + 1}} = \frac{1}{2}\sqrt{2}\,|y - x|.$$

Since the points on the curve are below the line $y = x$, we can replace $|y - x|$ by $x - y$. To simplify the work we drop the constant factor $\frac{1}{2}\sqrt{2}$.

$$f(x,y) = x - y, \qquad g(x,y) = x^3 - y^3 - 1$$

$$\nabla f = \mathbf{i} - \mathbf{j}, \qquad \nabla g = 3x^2\mathbf{i} - 3y^2\mathbf{j}.$$

We use the cross-product equation (16.8.4):

$$1\left(-3y^2\right) - \left(3x^2\right)\left(-1\right) = 0, \quad 3x^2 - 3y^2 = 0, \quad x = -y \ \ (x \neq y).$$

Now $g(x,y) = 0$ gives us

$$x^3 - (-x)^3 - 1 = 0, \quad 2x^3 = 1, \quad x = 2^{-1/3}.$$

The point is $\left(2^{-1/3}, -2^{-1/3}\right)$.

27.   It suffices to show that the square of the area is a maximum when $a = b = c$.

$$f(a,b,c) = s(s-a)(s-b)(s-c), \quad g(a,b,c) = a + b + c - 2s$$

$$\nabla f = -s(s-b)(s-c)\mathbf{i} - s(s-a)(s-c)\mathbf{j} - s(s-a)(s-b)\mathbf{k}, \quad \nabla g = \mathbf{i} + \mathbf{j} + \mathbf{k}.$$

$$\nabla f = \lambda \nabla g \implies -s(s-b)(s-c) = -s(s-a)(s-c) = -s(s-a)(s-b) = \lambda.$$

Thus, $s - b = s - a = s - c$ so that $a = b = c$. This gives us the maximum, as no minimum exists. [The area can be made arbitrarily small by taking a close to s.]

29.   (a)

$$f(x,y) = (xy)^{1/2}, \qquad g(x,y) = x + y - k, (x, y \geq 0, \ k \text{ a nonnegative constant}$$

$$\nabla f = \frac{y^{1/2}}{2x^{1/2}}\mathbf{i} + \frac{x^{1/2}}{2y^{1/2}}\mathbf{j} \qquad \nabla g = \mathbf{i} + \mathbf{j}.$$

$$\nabla f = \lambda \nabla g \implies \frac{y^{1/2}}{2x^{1/2}} = \lambda = \frac{x^{1/2}}{2y^{1/2}} \implies x = y = \frac{k}{2}.$$

Thus, the maximum value of f is: $f(k/2, \ k/2) = \dfrac{k}{2}$.

(b) For all x, y $(x, y \geq 0)$ we have

$$(xy)^{1/2} = f(x,y) \leq f(k/2, \ k/2) = \frac{k}{2} = \frac{x+y}{2}.$$

31.   Simply extend the arguments used in Exercises 29 and 30.

33. $$S(r,h) = 2\pi r^2 + 2\pi rh, \qquad g(r,h) = \pi r^2 h, \quad (V \text{ constant}$$

$$\nabla S = (4\pi r + 2\pi h)\mathbf{i} + 2\pi r\mathbf{j}, \qquad \nabla g = 2\pi rh\mathbf{i} + \pi r^2\mathbf{j}.$$

$$\nabla S = \lambda \nabla g \implies 4\pi r + 2\pi h = 2\pi rh\lambda, \quad 2\pi r = \pi r^2 \lambda \implies r = \frac{2}{\lambda}, \quad h = \frac{4}{\lambda}.$$

Now $\pi r^2 h = V, \implies \lambda = \sqrt[3]{\dfrac{16\pi}{V}} \implies r = \sqrt[3]{\dfrac{V}{2\pi}}, \quad h = \sqrt[3]{\dfrac{4V}{\pi}}.$

To minimize the surface area, take $r = \sqrt[3]{\dfrac{V}{2\pi}}$, and $h = \sqrt[3]{\dfrac{4V}{\pi}}$.

35. $S(l, w, h) = lw + 2(lh + wh)$, $\qquad\qquad$ $g(l, w, h) = lwh - 12$, $(V$ constant$)$

$\qquad \nabla S = (w + 2h)\mathbf{i} + (l + 2h)\mathbf{j} + 2(l + w)\mathbf{k}$, $\qquad\qquad$ $\nabla g = wh\,\mathbf{i} + lh\,\mathbf{j} + lw\,\mathbf{k}$.

$\qquad \nabla S = \lambda\nabla g \implies w + 2h = \lambda wh, \quad l + 2h = \lambda lh, \quad 2(l + w) = \lambda lw \implies w = l = 2h$.

Now $lwh = 12, \implies h = \sqrt[3]{3}, \ l = w = 2\sqrt[3]{3}$.

To minimize the surface area, take $l = w = 2\sqrt[3]{3}$ ft. and $h = \sqrt[3]{3}$ ft.

35. (a) $\qquad\qquad f(x, y, l) = xyl$, $\qquad\qquad\qquad\qquad$ $g(x, y, l) = 2x + 2y + l - 108$,

$\qquad\qquad \nabla f = yl\,\mathbf{i} + xl\,\mathbf{j} + xy\,w\mathbf{k}$, $\qquad\qquad$ $\nabla g = 2\mathbf{i} + 2\mathbf{j} + \mathbf{k}$.

$\qquad\qquad \nabla f = \lambda\nabla g \implies yl = 2\lambda, \quad xl = 2\lambda, \quad xy = \lambda \implies y = x$ and $l = 2x$.

Now $2x + 2y + l = 108, \implies x = 18$ and $l = 36$.

To maximize the volume, take $x = y = 18$ in. and $l = 36$ in.

$\qquad$ **(b)** $\qquad\qquad f(r, l) = \pi r^2 l$, $\qquad\qquad\qquad\qquad$ $g(r, l) = 2\pi r + l - 108$,

$\qquad\qquad\qquad \nabla f = 2\pi rl\,\mathbf{i} + \pi r^2\,\mathbf{j}$, $\qquad\qquad$ $\nabla g = 2\pi\,\mathbf{i} + \mathbf{j}$.

$\qquad\qquad\qquad \nabla f = \lambda\nabla g \implies 2\pi rl = 2\pi\lambda, \quad \pi r^2 = \lambda, \quad l = \pi r$.

Now $2\pi r + l = 108, \implies r = \dfrac{36}{\pi}$ and $l = 36$.

To maximize the volume, take $r = 36/\pi$ in. and $l = 36$ in.

39.   To simplify notation we set $x = Q_1, \quad y = Q_2, \quad z = Q_3$.

$\qquad f(x, y, z) = 2x + 8y + 24z$, $\qquad\qquad$ $g(x, y, z) = x^2 + 2y^2 + 4z^2 - 4{,}500{,}000{,}000$

$\qquad \nabla f = 2\mathbf{i} + 8\mathbf{j} + 24\mathbf{k}$, $\qquad\qquad\qquad$ $\nabla g = 2x\mathbf{i} + 4y\mathbf{j} + 8z\mathbf{k}$.

$\qquad\qquad \nabla f = \lambda\nabla g \implies 2 = 2\lambda x, \quad 8 = 4\lambda y, \quad 24 = 8\lambda z$.

Since $\lambda \neq 0$ here, we solve the equations for x, y, z:

$$x = \frac{1}{\lambda}, \quad y = \frac{2}{\lambda}, \quad z = \frac{3}{\lambda},$$

and substitute these results in $g(x, y, z) = 0$ to obtain

$$\frac{1}{\lambda^2} + 2\left(\frac{4}{\lambda^2}\right) + 4\left(\frac{9}{\lambda^2}\right) - 45 \times 10^8 = 0, \quad \frac{45}{\lambda^2} = 45 \times 10^8, \quad \lambda = \pm 10^{-4}.$$

Since x, y, z are non-negative, $\lambda = 10^{-4}$ and

$$x = 10^4 = Q_1, \quad y = 2 \times 10^4 = Q_2, \quad z = 3 \times 10^4 = Q_3.$$

41. $f(x, y, z) = xy^2z - x^2yz$, $g(x, y, z) = x^2 + y^2 - 1$, $h(x, y, z) = z^2 - x^2 - y^2$, $z \geq 0$

$\nabla f = (y^2z - 2xyz)\mathbf{i} + (2xyz - x^2z)\mathbf{j} + (xy^2 - x^2y)\mathbf{k}$, $\nabla g = 2x\mathbf{i} + 2y\mathbf{j}$, $\nabla h = -2x\mathbf{i} + 2y\mathbf{j} + 2z\mathbf{k}$.

$\nabla f = \lambda\nabla g + \mu\nabla h \implies$

$$y^2z - 2xyz = 2x\lambda - 2x\mu2x$$

$$2xyz - x^2z = 2y\lambda - 2y\mu$$

$$xy^2 - yx^2 = 2z\mu$$

Note first that $z = 1$. Setting $z = 1$ in the first two equations and adding, we get

$$y^2 - x^2 = 2x(\lambda - \mu) + 2y(\lambda - \mu) \quad\text{or}\quad (y + x)(y - x) = 2(y + x)(\lambda - \mu)$$

Therefore, either $y = -x$ or $\lambda - \mu = \dfrac{y - x}{2}$.

First, let $y = -x$.   Then it follows that $x = \dfrac{1}{\sqrt{2}}$, $y = -\dfrac{1}{\sqrt{2}}$, $z = 1$.

If $y \neq -x$, then $\lambda - \mu = \dfrac{y - x}{2}$ and

$$y^2 - 2xy = x(y - x)$$

$$2xy - x^2 = y(y - x)$$

Solving these equations simultaneously, using the fact that $x^2 + y^2 = 1$, we obtain

$$9x^4 - 9x^2 + 1 = 0 \implies x^2 = \frac{3 \pm \sqrt{5}}{6} \cong 0.873,\ 0.127$$

Now,

$$x^2 = 0.873 \implies y^2 = 0.127 \implies x = \pm0.934,\ y = \pm0.356;$$

$$x^2 = 0.127 \implies y^2 = 0.873 \implies x = \pm0.356,\ y = \pm0.934.$$

Clearly, the maximum value of f will occur when $x > 0$ and $y < 0$:

$$f\left(\frac{1}{\sqrt{2}}, -\frac{1}{\sqrt{2}}, 1\right) = \frac{\sqrt{2}}{2} \cong 0.707,$$

$$f(0.934, -0.356, 1) = f(0.356, -0.934, 1) \cong 0.429.$$

Therefore, the maximum value of f subject to the constraints is $\dfrac{\sqrt{2}}{2}$.

43. $f(x, y, z) = x^2 + y^2 + z^2$, $g(x, y, z) = x + y - z + 1 = 0$, $h(x, y, z) = x^2 + y^2 - z^2$

$\nabla f = 2x\mathbf{i} + 2y\mathbf{j} + 2z\,w\mathbf{k}$, $\nabla g = \mathbf{i} + \mathbf{j} - \mathbf{k}$, $\nabla h = 2x\mathbf{i} + 2y\mathbf{j} - 2z\,\mathbf{k}$.

$\nabla f = \lambda\nabla g + \mu\nabla h \implies 2x = \lambda + 2x\mu,\ \ 2y = \lambda + 2y\mu,\ \ 2z = -\lambda - 2z\mu$

Multiplying the first equation by y, the second equation by x and subtracting, yields

$$\lambda(y - x) = 0.$$

Now $\lambda = 0 \implies \mu = 1 \implies x = y = z = 0$. This is impossible since $x + y - z = -1$.

Therefore, we must have $y = x \implies z = \pm\sqrt{2}\,x$.

Substituting $y = x$, $z = \sqrt{2}\,x$ into the equation $x + y - z + 1 = 0$, we get

$$x = -1 - \frac{\sqrt{2}}{2} \implies y = -1 - \frac{\sqrt{2}}{2},\ z = -1 - \sqrt{2}$$

Substituting $y = x$, $z = -\sqrt{2}\,x$ into the equation $x + y - z + 1 = 0$, we get

$$x = -1 + \frac{\sqrt{2}}{2} \implies y = -1 + \frac{\sqrt{2}}{2},\ z = -1 + \sqrt{2}$$

Since

$$f\left(-1 - \frac{\sqrt{2}}{2}, -1 - \frac{\sqrt{2}}{2}, -1 - \sqrt{2}\right) = 6 + 4\sqrt{2}\ \text{ and}$$

$$f\left(-1 + \frac{\sqrt{2}}{2}, -1 + \frac{\sqrt{2}}{2}, -1 + \sqrt{2}\right) = 6 - 4\sqrt{2},$$

it follows that $\left(-1 + \frac{\sqrt{2}}{2}, -1 + \frac{\sqrt{2}}{2}, -1 + \sqrt{2}\right)$ is closest to the origin and

$\left(-1 - \frac{\sqrt{2}}{2}, -1 - \frac{\sqrt{2}}{2}, -1 - \sqrt{2}\right)$ is furthest from the origin.

SECTION 15.9

1. $df = \left(3x^2 y - 2xy^2\right)\Delta x + \left(x^3 - 2x^2 y\right)\Delta y$

3. $df = \left(\cos y + y \sin x\right)\Delta x - \left(x \sin y + \cos x\right)\Delta y$

5. $df = \Delta x - \left(\tan z\right)\Delta y - \left(y \sec^2 z\right)\Delta z$

7. $df = \dfrac{y(y^2 + z^2 - x^2)}{\left(x^2 + y^2 + z^2\right)^2}\Delta x + \dfrac{x(x^2 + z^2 - y^2)}{\left(x^2 + y^2 + z^2\right)^2}\Delta y - \dfrac{2xyz}{\left(x^2 + y^2 + z^2\right)^2}\Delta z$

9. $df = \left[\cos(x + y) + \cos(x - y)\right]\Delta x + \left[\cos(x + y) - \cos(x - y)\right]\Delta y$

11. $df = \left(y^2 z e^{xz} + \ln z\right)\Delta x + 2y e^{xz}\,\Delta y + \left(xy^2 e^{xz} + \dfrac{x}{z}\right)\Delta z$

13.
$$\Delta u = \left[(x + \Delta x)^2 - 3(x + \Delta x)(y + \Delta y) + 2(y + \Delta y)^2\right] - \left(x^2 - 3xy + 2y^2\right)$$

$$= \left[(1.7)^2 - 3(1.7)(-2.8) + 2(-2.8)^2\right] - \left(2^2 - 3(2)(-3) + 2(-3)^2\right)$$

$$= (2.89 + 14.28 + 15.68) - 40 = -7.15$$

$$du = (2x - 3y)\,\Delta x + (-3x + 4y)\,\Delta y$$

$$= (4 + 9)(-0.3) + (-18)(0.2) = -7.50$$

15.
$$\Delta u = \left[(x + \Delta x)^2(z + \Delta z) - 2(y + \Delta y)(z + \Delta z)^2 + 3(x + \Delta x)(y + \Delta y)(z + \Delta z)\right]$$

$$- \left(x^2 z - 2yz^2 + 3xyz\right)$$

$$= \left[(2.1)^2(2.8) - 2(1.3)(2.8)^2 + 3(2.1)(1.3)(2.8)\right] - \left[(2)^2 3 - 2(1)(3)^2 + 3(2)(1)(3)\right] = 2.896$$

$$du = (2xz + 3yz)\,\Delta x + \left(-2z^2 + 3xz\right)\,\Delta y + \left(x^2 - 4yz + 3xy\right)\,\Delta z$$

$$= [2(2)(3) + 3(1)(3)](0.1) + [-2(3)^2 + 3(2)(3)](0.3) + [2^2 - 4(1)(3) + 3(2)(1)](-0.2) = 2.5$$

17. $f(x, y) = x^{1/2} y^{1/4}; \quad x = 121, \quad y = 16, \quad \Delta x = 4, \quad \Delta y = 1$

$$f(x + \Delta x,\, y + \Delta y) \cong f(x, y) + df$$

$$= x^{1/2}\, y^{1/4} + \tfrac{1}{2} x^{-1/2}\, y^{1/4}\, \Delta x + \tfrac{1}{4} x^{1/2}\, y^{-3/4}\, \Delta y$$

$$\sqrt{125}\ \sqrt[4]{17} \cong \sqrt{121}\ \sqrt[4]{16} + \tfrac{1}{2}(121)^{-1/2}\,(16)^{1/4}\,(4) + \tfrac{1}{4}(121)^{1/2}\,(16)^{-3/4}\,(1)$$

$$= 11(2) + \tfrac{1}{2}\left(\tfrac{1}{11}\right)(2)(4) + \tfrac{1}{4}(11)\left(\tfrac{1}{8}\right)$$

$$= 22 + \tfrac{4}{11} + \tfrac{11}{32} = 22\tfrac{249}{352} \cong 22.71$$

19.
$$f(x, y) = \sin x \cos y; \quad x = \pi, \quad y = \frac{\pi}{4}, \quad \Delta x = -\frac{\pi}{7}, \quad \Delta y = -\frac{\pi}{20}$$

$$df = \cos x \cos y\,\Delta x - \sin x \sin y\,\Delta y$$

$$f(x + \Delta x,\, y + \Delta y) \cong f(x, y) + df$$

$$\sin \frac{6}{7}\pi \cos \frac{1}{5}\pi \cong \sin \pi \cos \frac{\pi}{4} + \left(\cos \pi \cos \frac{\pi}{4}\right)\left(-\frac{\pi}{7}\right) - \left(\sin \pi \sin \frac{\pi}{4}\right)\left(-\frac{\pi}{20}\right)$$

$$= 0 + \left(\frac{1}{2}\sqrt{2}\right)\left(\frac{\pi}{7}\right) + 0 = \frac{\pi\sqrt{2}}{14} \cong 0.32$$

21. $f(2.9, 0.01) \cong f(3, 0) + df,$ where df is to be evaluated at $x = 3,\ y = 0,\ \Delta x = -0.1,\ \Delta y = 0.01.$

$$df = \left(2xe^{xy} + x^2ye^{xy}\right)\Delta x + x^3 e^{xy}\,\Delta y = \left[2(3)e^0 + (2)^2(0)e^0\right](-0.1) + 3^3 e^0 (0.01) = -0.33$$

Thus, $f(2.9, .01) \cong 3^2 e^0 - 0.33 = 8.67$.

23. $f(2.94, 1.1, 0.92) \cong f(3, 1, 1) + df$,   where df is to be evaluated at $x = 3$, $y = 1$, $z = 1$,

$\Delta x = -0.06$, $\Delta y = 0.1$, $\Delta z = -0.08$

$$df = \tan^{-1} yz\,\Delta x + \frac{xz}{1 + y^2 z^2}\,\Delta y + \frac{xy}{1 + y^2 z^2}\,\Delta z = \frac{\pi}{4}(-0.06) + (1.5)(0.1) + (1.5)(-0.08) \cong -0.441$$

Thus, $f(2.94, 1.1, 0.92) \cong 3 - 0.441 = 2.559$

25. $df = \dfrac{\partial z}{\partial x}\,\Delta x + \dfrac{\partial z}{\partial y}\,\Delta y = \dfrac{2y}{(x+y)^2}\,\Delta x - \dfrac{2x}{(x+y)^2}\,\Delta y$

With $x = 4$, $y = 2$, $\Delta x = 0.1$, $\Delta y = 0.1$,   we get

$$df = \tfrac{4}{36}(0.1) - \tfrac{8}{36}(0.1) = -\tfrac{1}{90}.$$

The exact change is $\dfrac{4.1 - 2.1}{4.1 + 2.1} - \dfrac{4 - 2}{4 + 2} = \dfrac{2}{6.2} - \dfrac{1}{3} = -\dfrac{1}{93}.$

27. $S = 2\pi r^2 + 2\pi rh$; $r = 8$, $h = 12$, $\Delta r = -0.3$, $\Delta h = 0.2$

$$dS = \frac{\partial S}{\partial r}\,\Delta r + \frac{\partial S}{\partial h}\,\Delta h = (4\pi r + 2\pi h)\,\Delta r + (2\pi r)\,\Delta h$$

$$= 56\pi(-0.3) + 16\pi(0.2) = -13.6\pi.$$

The area decreases about 13.6π in.^{2}.

29. $S(9.98, 5.88, 4.08) \cong S(10, 6, 4) + dS = 248 + dS$,   where

$dS = (2w + 2h)\,\Delta l + (2l + 2h)\,\Delta w + (2l + 2w)\,\Delta h = 20(-0.02) + 28(-0.12) + 32(0.08) = -1.20$

Thus, $S(9.98, 5.88, 4.08) \cong 248 - 1.20 = 246.80$.

31.   (a) $dV = yx\,\Delta x + xz\,\Delta y + xy\,\Delta z = (8)(6)(0.02) + (12)(6)(-0.05) + (12)(8)(0.03) = 0.24$

(b) $\Delta V = (12.02)(7.95)(6.03) - (12)(8)(6) = 0.22077$

33. $T(P) - T(Q) \cong dT = (-2x + 2yz)\,\Delta x + (-2y + 2xz)\,\Delta y + (-2z + 2xy)\,\Delta z$

Letting $x = 1$, $y = 3$, $z = 4$, $\Delta x = 0.15$, $\Delta y = -0.10$, $\Delta z = 0.10$, we have

$$dT = (22)(0.15) + (2)(-0.10) + (-2)(0.10) = 2.9$$

35.   The area is given by $A = \frac{1}{2} x^2 \tan\theta$. The change in area is approximated by:

$$dA = x\tan\theta\,\Delta x + \tfrac{1}{2} x^2 \sec^2\theta\,\Delta\theta \cong 4(0.75)\,\Delta x + 8(1.5625)\,\Delta\theta = 3\,\Delta x + 12.5\,\Delta\theta$$

The area is more sensitive to a change in θ.

37. (a) $\pi r^2 h = \pi(r+\Delta r)^2(h+\Delta h) \implies \Delta h\,\dfrac{r^2 h}{(r+\Delta r)^2} - h = -\dfrac{(2r+\Delta r)h}{(r+\Delta r)^2}\,\Delta r.$

$$df = (2\pi rh)\,\Delta r + \pi r^2\,\Delta h, \qquad df = 0 \implies \Delta h = \dfrac{-2h}{r}\,\Delta r.$$

(b) $2\pi r^2 + 2\pi rh = 2\pi(r+\Delta r)^2 + 2\pi(r+\Delta r)(h+\Delta h).$

Solving for Δh,

$$\Delta h = \dfrac{r^2 + rh - (r+\Delta r)^2}{r+\Delta r} - h = -\dfrac{(2r+h+\Delta r)}{r+\Delta r}\,\Delta r.$$

$$df = (4\pi r + 2\pi h)\,\Delta r + 2\pi r\,\Delta h, \qquad df = 0 \implies \Delta h = -\left(\dfrac{2r+h}{r}\right)\Delta r.$$

39. (a) $c(x,y) = \sqrt{x^2+y^2};\ x=5, y=12, \Delta x = \pm 1.5, \Delta y = \pm 1.5$

$$dc = \frac{\partial c}{\partial x}\,\Delta x + \frac{\partial c}{\partial y}\,\Delta y = \frac{x}{\sqrt{x^2+y^2}}\,\Delta x + \frac{y}{\sqrt{x^2+y^2}}\,\Delta y$$

$$= \frac{5}{13}(\pm 1.5) + \frac{12}{13}(\pm 1.5) \cong \pm 1.962$$

The maximum possible error in the value of the hypotenuse is 1.962 cm.

(b) $A(x,y) = \tfrac{1}{2}xy;\ x=5, y=12, \Delta x = \pm 1.5, \Delta y = \pm 1.5$

$$dA = \frac{\partial A}{\partial x}\,\Delta x + \frac{\partial A}{\partial y}\,\Delta y = \tfrac{1}{2}\,y\,\Delta x + \tfrac{1}{2}\,y\,\Delta y$$

$$= \tfrac{1}{2}(12)(\pm 1.5) + \tfrac{1}{2}(5)(\pm 1.5) \cong \pm 12.75$$

The maximum possible error in the value of the area is 12.75 cm^2.

41. $s = \dfrac{A}{A-W};\ A=9,\ W=5,\ \Delta A = \pm 0.01,\ \Delta W = \pm 0.02$

$$ds = \frac{\partial s}{\partial A}\,\Delta A + \frac{\partial s}{\partial W}\,\Delta W = \frac{-W}{(A-W)^2}\,\Delta A + \frac{A}{(A-W)^2}\,\Delta W$$

$$= -\frac{5}{16}(\pm 0.01) + \frac{9}{16}(\pm 0.02) \cong \pm 0.014$$

The maximum possible error in the value of s is 0.014 lbs; $2.23 \leq s + \Delta s \leq 2.27$

SECTION 15.10

1. $\dfrac{\partial f}{\partial x} = xy^2,\quad f(x,y) = \tfrac{1}{2}x^2y^2 + \phi(y),\quad \dfrac{\partial f}{\partial y} = x^2y + \phi'(y) = x^2y.$

Thus, $\phi'(y) = 0,\ \phi(y) = C,$ and $f(x,y) = \tfrac{1}{2}x^2y^2 + C.$

3. $\dfrac{\partial f}{\partial x} = y,\quad f(x,y) = xy + \phi(y),\quad \dfrac{\partial f}{\partial y} = x + \phi'(y) = x.$

Thus, $\phi'(y) = 0$, $\phi(y) = C$, and $f(x,y) = xy + C$.

5. No; $\dfrac{\partial}{\partial y}\left(y^3 + x\right) = 3y^2$ whereas $\dfrac{\partial}{\partial x}\left(x^2 + y\right) = 2x$.

7. $\dfrac{\partial f}{\partial x} = \cos x - y\sin x$, $f(x,y) = \sin x + y\cos x + \phi(y)$, $\dfrac{\partial f}{\partial y} = \cos x + \phi'(y) = \cos x$.

Thus, $\phi'(y) = 0$, $\phi(y) = C$, and $f(x,y) = \sin x + y\cos x + C$.

9. $\dfrac{\partial f}{\partial x} = e^x \cos y^2$, $f(x,y) = e^x \cos y^2 + \phi(y)$, $\dfrac{\partial f}{\partial y} = -2ye^x \sin y^2 + \phi'(y) = -2ye^x \sin y^2$.

Thus, $\phi'(y) = 0$, $\phi(y) = C$, and $f(x,y) = e^x \cos y^2 + C$.

11. $\dfrac{\partial f}{\partial y} = xe^x - e^{-y}$, $f(x,y) = xye^x + e^{-y} + \phi(x)$, $\dfrac{\partial f}{\partial x} = ye^x + xye^x + \phi'(x) = ye^x(1+x)$.

Thus, $\phi'(x) = 0$, $\phi(x) = C$, and $f(x,y) = xye^x + e^{-y} + C$.

13. No; $\dfrac{\partial}{\partial y}\left(xe^{xy} + x^2\right) = x^2 e^{xy}$ whereas $\dfrac{\partial}{\partial x}\left(ye^{xy} - 2y\right) = y^2 e^{xy}$

15. $\dfrac{\partial}{\partial x} = 1 + y^2 + xy^2$, $f(x,y) = x + xy^2 + \tfrac{1}{2}x^2 y^2 + \phi(y)$, $\dfrac{\partial}{\partial y} = 2xy + x^2 y + \phi'(y) = x^2 y + y + 2xy + 1$.

Thus, $\phi'(y) = y + 1$, $\phi(y) = \tfrac{1}{2}y^2 + y + C$ and $f(x,y) = x + xy^2 + \tfrac{1}{2}x^2 y^2 + \tfrac{1}{2}y^2 + y + C$.

17. $\dfrac{\partial f}{\partial x} = \dfrac{x}{\sqrt{x^2 + y^2}}$, $f(x,y) = \sqrt{x^2 + y^2} + \phi(y)$, $\dfrac{\partial f}{\partial y} = \dfrac{y}{\sqrt{x^2 + y^2}} + \phi'(y) = \dfrac{y}{\sqrt{x^2 + y^2}}$.

Thus, $\phi'(y) = 0$, $\phi(y) = C$, and $f(x,y) = \sqrt{x^2 + y^2} + C$.

19. $\dfrac{\partial f}{\partial x} = x^2 \sin^{-1} y$, $f(x,y) = \tfrac{1}{3}x^3 \sin^{-1} y + \phi(y)$, $\dfrac{\partial f}{\partial y} = \dfrac{x^3}{3\sqrt{1-y^2}} + \phi'(y) = \dfrac{x^3}{3\sqrt{1-y^2}} - \ln y$.

Thus, $\phi'(y) = -\ln y$, $\phi(y) = y - y\ln y + C$, and

$$f(x,y) = \frac{1}{3}\sin^{-1} y + y - y\ln y + C.$$

21. $\dfrac{\partial f}{\partial x} = f(x,y)$, $\dfrac{\partial f/\partial x}{f(x,y)} = 1$, $\ln f(x,y) = x + \phi(y)$, $\dfrac{\partial f/\partial y}{f(x,y)} = 0 + \phi'(y)$, $\dfrac{\partial f}{\partial y} = f(x,y)$.

Thus, $\phi'(y) = 1$, $\phi(y) = y + K$, and $f(x,y) = e^{x+y+K} = Ce^{x+y}$.

23. (a) $P = 2x$, $Q = z$, $R = y$; $\dfrac{\partial P}{\partial y} = 0 = \dfrac{\partial Q}{\partial x}$, $\dfrac{\partial P}{\partial z} = 0 = \dfrac{\partial R}{\partial x}$, $\dfrac{\partial Q}{\partial z} = 1 = \dfrac{\partial R}{\partial y}$

(b), (c), and (d)

$$\frac{\partial f}{\partial x} = 2x, \quad f(x,y,z) = x^2 + g(y,z).$$

$$\frac{\partial f}{\partial y} = 0 + \frac{\partial g}{\partial y} \quad \text{with} \quad \frac{\partial f}{\partial y} = z \implies \frac{\partial g}{\partial y} = z.$$

Then,

$$g(y,z) = yz + h(z),$$

$$f(x,y,z) = x^2 + yz + h(z),$$

$$\frac{\partial f}{\partial z} = 0 + y + h'(z) \quad \text{and} \quad \frac{\partial f}{\partial z} = y \implies h'(z) = 0.$$

Thus, $h(z) = C$ and $f(x,y,z) = x^2 + yz + C$.

25. The function is a gradient by the test stated before Exercise 23.

Take $P = 2x + y$, $Q = 2y + x + z$, $R = y - 2z$. Then

$$\frac{\partial P}{\partial y} = 1 = \frac{\partial Q}{\partial x}, \quad \frac{\partial P}{\partial z} = 0 = \frac{\partial R}{\partial x}, \quad \frac{\partial Q}{\partial z} = 1 = \frac{\partial R}{\partial y}.$$

Next, we find f where $\nabla f = P\mathbf{i} + Q\mathbf{j} + R\mathbf{k}$.

$$\frac{\partial f}{\partial x} = 2x + y,$$

$$f(x,y,z) = x^2 + xy + g(y,z).$$

$$\frac{\partial f}{\partial y} = x + \frac{\partial g}{\partial y} \quad \text{with} \quad \frac{\partial f}{\partial y} = 2y + x + z \implies \frac{\partial g}{\partial y} = 2y + z.$$

Then,

$$g(y,z) = y^2 + yz + h(z),$$

$$f(x,y,z) = x^2 + xy + y^2 + yz + h(z).$$

$$\frac{\partial f}{\partial z} = y + h'(z) = y - 2z \implies h'(z) = -2z.$$

Thus, $h(z) = -z^2 + C$ and $f(x,y,z) = x^2 + xy + y^2 + yz - z^2 + C$.

27. The function is a gradient by the test stated before Exercise 23.

Take $P = y^2 z^3 + 1$, $Q = 2xyz^3 + y$, $R = 3xy^2 z^2 + 1$. Then

$$\frac{\partial P}{\partial y} = 2yz^3 = \frac{\partial Q}{\partial x}, \quad \frac{\partial P}{\partial z} = 3y^2 z^2 = \frac{\partial R}{\partial x}, \quad \frac{\partial Q}{\partial z} = 6xyz^2 = \frac{\partial R}{\partial y}.$$

Next, we find f where $\nabla f = P\mathbf{i} + Q\mathbf{j} + R\mathbf{k}$.

$$\frac{\partial f}{\partial x} = y^2 z^3 + 1,$$

$$f(x, y, z) = xy^2 z^3 + x + g(y, z).$$

$$\frac{\partial f}{\partial y} = 2xyz^3 \frac{\partial g}{\partial y} \quad \text{with} \quad \frac{\partial f}{\partial y} = 2xyz^3 + y \quad \Longrightarrow \quad \frac{\partial g}{\partial y} = y.$$

Then,

$$g(y, z) = \tfrac{1}{2} y^2 + h(z),$$

$$f(x, y, z) = xy^2 z^3 + x + \tfrac{1}{2} y^2 + h(z).$$

$$\frac{\partial f}{\partial z} = 3xy^2 z^2 + h'(z) = 3xy^2 z^2 + 1 \quad \Longrightarrow \quad h'(z) = 1.$$

Thus, $h(z) = z + C$ and $f(x, y, z) = xy^2 z^3 + x + \tfrac{1}{2} y^2 + z + C.$

29. $\mathbf{F}(\mathbf{r}) = \nabla \left(\dfrac{GmM}{r} \right)$

PROJECTS AND EXPLORATIONS

15.1.   (a)

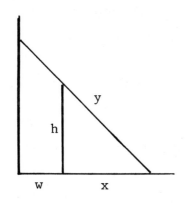

Minimize $L = \sqrt{(w + x)^2 + y^2}$

subject to $\dfrac{y}{w + x} = \dfrac{h}{x}$

(b) $L(x) = (w + x)\sqrt{1 + (h/x)^2}, \qquad 0 < x < \infty.$

$$\frac{dL}{dx} = \left(1 - \frac{wh^2}{x^3} \right) \left(1 + \frac{h^2}{x^2} \right)^{-1/2}; \qquad \frac{dL}{dx} = 0 \quad \Rightarrow \quad x = \sqrt[3]{wh^2}$$

It is easy to verify that L has a minimum at $x = \sqrt[3]{wh^2}$

When $h = 6$ and $w = 4$, $x = 5.24148$ and $L = 14.04696$.

(c) Let $f(x, y) = \sqrt{(4 + x)^2 + y^2}$, $g(x, y) = \dfrac{y}{4 + x} - \dfrac{6}{x}$, and solve

$$\nabla f(x, y) = \lambda \nabla g(x, y)$$

Numerical solutions are: $x = 5.24148$, $y = 10.57886$, $\lambda = 0.14368$, $L = 14.04696$.

(d) Minimize $L = L_1 + L_2 = \sqrt{(u+x)^2 + y^2} + \sqrt{(v+w-x)^2 + y^2}$

subject to: $\dfrac{h_1}{u} = \dfrac{y}{u+x}$, $\dfrac{h_2}{v} = \dfrac{y}{w+v-x}$, where $0 \le x \le w$, $y > \max\{h_1, h_2\}$

If x is fixed and y varies, then

$$\lim_{y \to \max\{h_1,h_2\}} L = \infty \quad \text{and} \quad \lim_{y \to \infty} L = \infty.$$

Thus, for a fixed x, there must be a value of y which minimizes L..

Since $0 \le x \le w$ and all functions are continuous, there must be a minimum of the minimums.

(e) $L_1 = \sqrt{\left(\dfrac{x}{y-h_1}\right)^2 + 1}$, $L_2 = \sqrt{\left(\dfrac{w-x}{y-h_2}\right)^2 + 1}$

(f)

x	y	L_1	L_2	L
5	8	15.5492	60.5300	76.0802
7	10	17.2047	34.0037	51.2083
10	12	20.9255	23.3238	44.2493
12	15	23.4308	20.0693	43.5001

(g) Software which solves nonlinear systems of equations can be used to solve this system:

$$x = 12.120172, \ y = 13.566945, \ L = 43.091936$$

A plot of the surface $L(x,y)$ indicates that this is the absolute minimum.

15.3. (a) The cost of doing no sorting is 0; the cost of sorting everything is infinite.

$$C(\gamma) = T + c_4 \frac{\gamma}{1-\gamma} + c_3(1-\gamma) - p\gamma$$

where T depends on the location of the recycling center which depends on γ.

(b)
$$T(\gamma) = \begin{cases} 2 + 2c_2^2(1-\gamma)^2 + c_2(1-\gamma)(3 - c_2 + 4c_2\gamma), & c_2(1-\gamma) \le \frac{2}{3} \\ 6 + \frac{1}{2}c_2^2(1-\gamma)^2 + c_2(1-\gamma)(1 - c_2 + c_2\gamma), & c_2(1-\gamma) > \frac{2}{3}. \end{cases}$$

(i) $c_2 = 0.5$, $c_3 = 1$, $c_4 = 1$, with $p = 1$; $\gamma = 0.500001$, $C = 3.875$.

(ii) $c_2 = 0.5$, $c_3 = 0.5$, $c_4 = 1$, with $p = 0.5$; $\gamma = 0.42933$, $C = 3.84183$.

(iii) $c_2 = 0.5$, $c_3 = 1$, $c_4 = 0.5$, with $p = 0.25$; $\gamma = 0.60150$, $C = 3.67998$.

(iv) $c_2 = 0.5$, $c_3 = 0.5$, $c_4 = 0.5$, with $p = 0.5$; $\gamma = 0.58579$, $C = 3.32843$.

(v) $c_2 = 0.25$, $c_3 = 1$, $c_4 = 1$, with $p = 0.1$; $\gamma = 0.29738$, $C = 3.68480$.

(c)-(d) The recycling center is at the point $(1, 3)$. We give the results for (b) (i).

$$c_2 = 0.5, \quad c_3 = 1, \quad c_4 = 1, \quad \text{with } p = 1.$$

$$\gamma = 0.46458, \quad C = 5.74166; \text{ the cost is up } 1.866$$

$$c_2 = 0.51, \quad c_3 = 1, \quad c_4 = 1, \quad \text{with } p = 1.$$

$$\gamma = 0.46775, \quad C = 5.75766; \quad \frac{\partial \gamma}{\partial c_2} \cong 0.2275, \quad \frac{\partial C}{\partial c_2} \cong 1.6001.$$

If the cost of driving to the landfill goes up, the percentage to be recycled should also go up.

$$c_2 = 0.5, \quad c_3 = 0.99, \quad c_4 = 1, \quad \text{with } p = 1.$$

$$\gamma = 0.46471, \quad C = 5.73631; \quad \frac{\partial \gamma}{\partial c_3} \cong 0.0764, \quad \frac{\partial C}{\partial c_2} \cong 0.5349.$$

If the cost of dumping decreases, then so does the percentage recycled and the costs. However, note that this is only one-third (approx.) the rate of change for the same change in c_2.

$$c_2 = 0.5, \quad c_3 = 1, \quad c_4 = 0.99, \quad \text{with } p = 1.$$

$$\gamma = 0.46816, \quad C = 5.73290; \quad \frac{\partial \gamma}{\partial c_2} \cong -0.2678, \quad \frac{\partial C}{\partial c_2} \cong 0.8755.$$

If the cost of sorting decreases, then we increase the amount sorted and decrease total costs.

CHAPTER 16

SECTION 16.1

1. $\displaystyle\sum_{i=1}^{3}\sum_{j=1}^{3} 2^{i-1}3^{j+1} = \left(\sum_{i=1}^{3} 2^{i-1}\right)\left(\sum_{j=1}^{3} 3^{j+1}\right) = (1+2+4)(9+27+81) = 819$

3. $\displaystyle\sum_{i=1}^{4}\sum_{j=1}^{3}(i^2+3i)(j-2) = \left[\sum_{i=1}^{4}(i^2+31)\right]\left[\sum_{j=1}^{3}(j-2)\right] = (4+10+18+28)(-1+0+1) = 0$

5. $\displaystyle\sum_{i=1}^{m}\Delta x_i = \Delta x_1 + \Delta x_2 + \cdots + \Delta x_n = (x_1-x_0)+(x_2-x_1)+\cdots+(x_n-x_{n-1})$

$$= x_n - x_0 = a_2 - a_1$$

7. $\displaystyle\sum_{i=1}^{m}\sum_{j=1}^{n}\Delta x_i\,\Delta y_j = \left(\sum_{i=1}^{m}\Delta x_i\right)\left(\sum_{j=1}^{n}\Delta y_j\right) = (a_2-a_1)(b_2-b_1)$

9. $\displaystyle\sum_{i=1}^{m}(x_i+x_{i-1})\,\Delta x_i = \sum_{i=1}^{m}(x_i+x_{i-1})(x_i-x_{i-1}) = \sum_{i=1}^{m}\left(x_i^2 - x_{i-1}^2\right)$

$$= x_m^2 - x_0^2 = a_2^2 - a_1^2$$

11. $\displaystyle\sum_{i=1}^{m}\sum_{j=1}^{n}(x_i+x_{i-1})\,\Delta x_i\,\Delta y_j = \left(\sum_{i=1}^{m}(x_i+x_{i-1})\,\Delta x_i\right)\left(\sum_{j=1}^{n}\Delta y_j\right)$

(Exercise 9) $\longrightarrow$

$$= \left(a_2^2 - a_1^2\right)(b_2 - b_1)$$

13. $\displaystyle\sum_{i=1}^{m}\sum_{j=1}^{n}(2\Delta x_i - 3\Delta y_j) = 2\left(\sum_{i=1}^{m}\Delta x_i\right)\left(\sum_{j=1}^{n}1\right) - 3\left(\sum_{i=1}^{m}1\right)\left(\sum_{j=1}^{n}\Delta y_j\right)$

$$= 2n\,(a_2 - a_1) - 3m\,(b_2 - b_1)$$

15. $\displaystyle\sum_{i=1}^{m}\sum_{j=1}^{n}\sum_{k=1}^{q}\Delta x_i\,\Delta y_j\,\Delta z_k = \left(\sum_{i=1}^{m}\Delta x_i\right)\left(\sum_{j=1}^{n}\Delta y_j\right)\left(\sum_{k=1}^{q}\Delta z_k\right)$

$$= (a_2 - a_1)(b_2 - b_1)(c_2 - c_1)$$

17. $\displaystyle\sum_{i=1}^{n}\sum_{j=1}^{n}\sum_{k=1}^{n}\delta_{ijk}a_{ijk}=a_{111}+a_{222}+\cdots+a_{nnn}=\sum_{p=1}^{n}a_{ppp}$

19. $\displaystyle\sum_{i=1}^{m}\sum_{j=1}^{n}\alpha a_{ij}=\sum_{i=1}^{m}\alpha\left(\sum_{j=1}^{n}a_{ij}\right)=\alpha\sum_{i=1}^{m}\sum_{j=1}^{n}a_{ij}$

SECTION 16.2

1. $\quad L_f(P)=2\frac{1}{4},\quad U_f(P)=5\frac{3}{4}$

3. (a) $\quad L_f(P)=\displaystyle\sum_{i=1}^{m}\sum_{j=1}^{n}(x_{i-1}+2y_{j-1})\,\Delta x_i\,\Delta y_j,\quad U_f(P)=\sum_{i=1}^{m}\sum_{j=1}^{n}(x_i+2y_j)\,\Delta x_i\,\Delta y_j$

(b) $\quad L_f(P)\le\displaystyle\sum_{i=1}^{m}\sum_{j=1}^{n}\left[\frac{x_{i-1}+x_i}{2}+2\left(\frac{y_{j-1}+y_j}{2}\right)\right]\Delta x_i\,\Delta y_j\le U_f(P).$

The middle expression can be written

$$\sum_{i=1}^{m}\sum_{j=1}^{n}\frac{1}{2}\left(x_i{}^2-x_{i-1}^2\right)\Delta y_j+\sum_{i=1}^{m}\sum_{j=1}^{n}\left(y_j{}^2-y_{j-1}^2\right)\Delta x_i.$$

The first double sum reduces to

$$\sum_{i=1}^{m}\sum_{j=1}^{n}\frac{1}{2}\left(x_i{}^2-x_{i-1}^2\right)\Delta y_j=\frac{1}{2}\left(\sum_{i=1}^{m}\left(x_i{}^2-x_{i-1}^2\right)\right)\left(\sum_{j=1}^{n}\Delta y_j\right)=\frac{1}{2}\left(4-0\right)\left(1-0\right)=2.$$

In like manner the second double sum also reduces to 2. Thus, $I=4$; the volume of the prism bounded above by the plane $z=x+2y$ and below by R.

5. $\quad L_f(P)=-7/24,\quad U_f(P)=7/24$

7. (a) $\quad L_f(P)=\displaystyle\sum_{i=1}^{m}\sum_{j=1}^{n}(4x_{i-1}\,y_{j-1})\,\Delta x_i\,\Delta y_j,\quad U_f(P)=\sum_{i=1}^{m}\sum_{j=1}^{n}(4x_i\,y_j)\,\Delta x_i\,\Delta y_j$

(b) $\quad L_f(P)\le\displaystyle\sum_{i=1}^{m}\sum_{j=1}^{n}(x_i+x_{i-1})(y_j+y_{j-1})\,\Delta x_1\,\Delta y_j\le U_f(P).$

The middle expression can be written

$$\sum_{i=1}^{m}\sum_{j=1}^{n}\left(x_i{}^2 - x_{i-1}^2\right)\left(y_j{}^2 - y_{j-1}^2\right) = \left(\sum_{i=1}^{m} x_i{}^2 - x_{i-1}^2\right)\left(\sum_{j=1}^{n} y_j{}^2 - y_{j-1}^2\right)$$

by (16.1.5)

$$= \left(b^2 - 0^2\right)\left(d^2 - 0^2\right) = b^2 d^2.$$

It follows that $I = b^2 d^2$.

9.   (a) $L_f(P) = \displaystyle\sum_{i=1}^{m}\sum_{j=1}^{n} 3\left(x_{i-1}^2 - y_j{}^2\right)\Delta x_i\,\Delta y_j,\quad U_f(P) = \displaystyle\sum_{i=1}^{m}\sum_{j=1}^{n} 3\left(x_i{}^2 - y_{j-1}^2\right)\Delta x_i\,\Delta y_j$

(b) $L_f(P) \le \displaystyle\sum_{i=1}^{m}\sum_{j=1}^{n}\left[\left(x_i{}^2 + x_i x_{i-1} + x_{i-1}^2\right) - \left(y_j{}^2 + y_j y_{j-1} + y_{j-1}^2\right)\right]\Delta x_i\,\Delta y_j \le U_f(P).$

Since in general $\left(A^2 + AB + B^2\right)\left(A - B\right) = A^3 - B^3,$ the middle expression can be written

$$\sum_{i=1}^{m}\sum_{j=1}^{n}\left(x_i{}^3 - x_{i-1}^3\right)\Delta y_j - \sum_{i=1}^{m}\sum_{j=1}^{n}\left(y_j{}^3 - y_{j-1}^3\right)\Delta x_i,$$

which reduces to

$$\left(\sum_{i=1}^{m} x_i{}^3 - x_{i-1}^3\right)\left(\sum_{j=1}^{n}\Delta y_j\right) - \left(\sum_{i=1}^{m}\Delta x_i\right)\left(\sum_{j=1}^{n} y_j{}^3 - y_{j-1}^3\right).$$

This can be evaluated as $b^3 d - b d^3 = bd\left(b^2 - d^2\right).$ It follows that $I = bd\left(b^2 - d^2\right).$

11. $0 \le \sin(x + y) \le 1$ for all $(x, y) \in R$. Thus, $0 \le \displaystyle\iint_R \sin(x + y)\,dx\,dy \le \iint_R dx\,dy = 1$

SECTION 16.3

1. $\displaystyle\iint_\Omega dx\,dy = \int_a^b \phi(x)\,dx$

3.   Suppose $f(x_0, y_0) \ne 0$. Assume $f(x_0, y_0) > 0$. Since f is continuous, there exists a disc Ω_ϵ with radius ϵ centered at (x_0, y_0) such that $f(x, y) > 0$ on Ω_ϵ. Let R be a rectangle contained in Ω_ϵ. Then $\displaystyle\iint_R f(x, y)\,dx\,dy > 0$, which contradicts the hypothesis.

5.   By Exercise 7, Section 16.2, $\displaystyle\iint_R 4xy\,dx\,dy = 2^2 3^2 = 36.$ Thus

$$f_{avg} = \frac{1}{\text{area}(R)}\iint_R 4xy\,dx\,dy = \frac{1}{6}\,(36) = 6$$

7.   By Theorem 16.3.5, there exists a point $(x_1, y_1) \in D_r$ such that

$$\iint\limits_{D_r} f(x,y)\,dx\,dy = f(x_1,y_1)\iint\limits_{R} dx\,dy = f(x_1,y_1)\pi r^2 \quad\Longrightarrow\quad f(x_1,y_1) = \frac{1}{\pi r^2}\iint\limits_{D_r} f(x,y)\,dx\,dy$$

As $r \to 0$, $(x_1,y_1) \to (x_0,y_0)$ and $f(x_1,y_1) \to f(x_0,y_0)$ since f is continuous.

The result follows.

SECTION 16.4

1. $\displaystyle\int_0^1\int_0^3 x^2\,dy\,dx = \int_0^1 3x^2\,dx = 1$

3. $\displaystyle\int_0^1\int_0^3 xy^2\,dy\,dx = \int_0^1 x\left[\frac{1}{3}y^3\right]_0^3 dx = \int_0^1 9x\,dx = \frac{9}{2}$

5. $\displaystyle\int_0^1\int_0^x xy^3\,dy\,dx = \int_0^1 x\left[\frac{1}{4}y^4\right]_0^x dx = \int_0^1 \frac{1}{4}x^5\,dx = \frac{1}{24}$

7. $\displaystyle\int_0^{\pi/2}\int_0^{\pi/2}\sin(x+y)\,dy\,dx = \int_0^{\pi/2}[-\cos(x+y)]_0^{\pi/2}\,dx = \int_0^{\pi/2}\left[\cos x - \cos\left(x+\frac{\pi}{2}\right)\right]dx = 2$

9. $\displaystyle\int_0^{\pi/2}\int_0^{\pi/2}(1+xy)\,dy\,dx = \int_0^{\pi/2}\left[y+\frac{1}{2}xy^2\right]_0^{\pi/2} dx = \int_0^{\pi/2}\left(\frac{1}{2}\pi+\frac{1}{8}\pi^2 x\right)dx = \frac{1}{4}\pi^2 + \frac{1}{64}\pi^4$

11. $\displaystyle\int_0^1\int_{y^2}^y \sqrt{xy}\,dx\,dy = \int_0^1 \sqrt{y}\left[\frac{2}{3}x^{3/2}\right]_{y^2}^y dy = \int_0^1 \frac{2}{3}\left(y^2 - y^{7/2}\right)dy = \frac{2}{27}$

13. $\displaystyle\int_{-2}^2\int_{\frac{1}{2}y^2}^{4-\frac{1}{2}y^2}\left(4-y^2\right)dx\,dy = \int_{-2}^2\left(4-y^2\right)\left[\left(4-\frac{1}{2}y^2\right)-\left(\frac{1}{2}y^2\right)\right]dy$

$$= 2\int_0^2\left(16-8y^2+y^4\right)dy = \frac{512}{15}$$

15. 0 by symmetry (integrand odd in y, Ω symmetric about x-axis)

17. $\displaystyle\int_0^2\int_0^{x/2} e^{x^2}\,dy\,dx = \int_0^2 \frac{1}{2}xe^{x^2}\,dx = \left[\frac{1}{4}e^{x^2}\right]_0^2 = \frac{1}{4}\left(e^4-1\right)$

19.

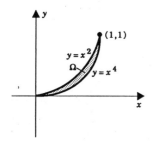

$$\int_0^1\int_{y^{1/2}}^{y^{1/4}} f(x,y)\,dx\,dy$$

21.

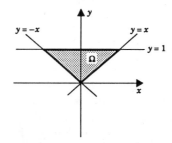

$$\int_{-1}^0\int_{-x}^1 f(x,y)\,dy\,dx + \int_0^1\int_x^1 f(x,y)\,dy\,dx$$

23.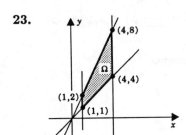

$$\int_1^2 \int_1^y f(x,y)\,dx\,dy + \int_2^4 \int_{y/2}^y f(x,y)\,dx\,dy$$

$$+ \int_4^8 \int_{y/2}^4 f(x,y)\,dx\,dy$$

25. $\displaystyle \int_{-2}^4 \int_{1/4x^2}^{\frac{1}{2}x+2} dy\,dx = \int_{-2}^4 \left[\frac{1}{2}x + 2 - \frac{1}{4}x^2\right] dx = 9$

27. $\displaystyle \int_0^{1/4} \int_{2y^{3/2}}^y dx\,dy = \int_0^{1/4} \left[y - 2y^{3/2}\right] dy = \frac{1}{160}$

29.

$$\int_0^1 \int_0^{y^2} \sin\left(\frac{y^3+1}{2}\right) dx\,dy = \int_0^1 y^2 \sin\left(\frac{y^3+1}{2}\right) dy$$

$$= \left[-\frac{2}{3}\cos\left(\frac{y^3+1}{2}\right)\right]_0^1$$

$$= \tfrac{2}{3}\left(\cos\tfrac{1}{2} - \cos 1\right)$$

31.

$$\int_0^{\ln 2} \int_{e^x}^2 e^{-x}\,dy\,dx = \int_0^{\ln 2} e^{-x}\left(2 - e^x\right) dx$$

$$= \left[-2e^{-x} - x\right]_0^{\ln 2} = 1 - \ln 2$$

33. $\displaystyle \int_1^2 \int_{y-1}^{2/y} dx\,dy = \int_1^2 \left[\frac{2}{y} - (y-1)\right] dy = \ln 4 - \frac{1}{2}$

35. $\displaystyle \int_0^2 \int_0^{3-\frac{3}{2}x} \left(4 - 2x - \frac{4}{3}y\right) dy\,dx = \int_0^3 \int_0^{2-\frac{2}{3}y} \left(4 - 2x - \frac{4}{3}y\right) dx\,dy = 4$

37. $\displaystyle \int_0^2 \int_0^{1-\frac{1}{2}x} x^3 y\,dy\,dx = \int_0^2 \int_0^{2-2y} x^3 y\,dx\,dy = \frac{2}{15}$

39.
$$\int_0^2 \int_{-\sqrt{2x-x^2}}^{\sqrt{2x-x^2}} (2x+1)\, dy\, dx = \int_{-1}^1 \int_{1-\sqrt{1-y^2}}^{1+\sqrt{1-y^2}} (2x+1)\, dx\, dy$$

$$= \int_{-1}^1 \left[x^2 + x \right]_{1-\sqrt{1-y^2}}^{1+\sqrt{1-y^2}} dy$$

$$= 6 \int_{-1}^1 \sqrt{1-y^2}\, dy = 6 \left(\frac{\pi}{2} \right) = 3\pi$$

41.
$$\int_0^1 \int_0^{1-x} (x^2+y^2)\, dy\, dx = \int_0^1 \left(2x^2 - \frac{4}{3}x^3 - x + \frac{1}{3} \right) dx = \frac{1}{6}$$

43.
$$\int_0^1 \int_{x^2}^x (x^2+3y^2)\, dy\, dx = \int_0^1 (2x^3 - x^4 - x^6)\, dx = \frac{11}{70}$$

45.
$$\int_0^a \int_0^{\sqrt{a^2-x^2}} \sqrt{a^2-x^2}\, dy\, dx = \int_0^a (a^2-x^2)\, dx\, dy = \frac{2}{3} a^2$$

47.
$$\int_0^1 \int_y^1 e^{y/x}\, dx\, dy = \int_0^1 \int_0^x e^{y/x}\, dy\, dx = \int_0^1 \left[x e^{y/x} \right]_0^x dx = \int_0^1 x(e-1)\, dx = \frac{1}{2}(e-1)$$

49.
$$\int_0^1 \int_x^1 x^2 e^{y^4}\, dy\, dx = \int_0^1 \int_0^y x^2 e^{y^4}\, dx\, dy = \int_0^1 \left[\frac{1}{3}x^3 e^{y^4} \right]_0^y dy = \frac{1}{3} \int_0^1 y^3 e^{y^4}\, dy = \frac{1}{12}(e-1)$$

51.
$$f_{avg} = \frac{1}{8} \int_{-1}^1 \int_0^4 x^2 y\, dy\, dx = \frac{1}{8} \int_{-1}^1 8x^2\, dx = \int_{-1}^1 x^2\, dx = \frac{2}{3}$$

53.
$$f_{avg} = \frac{1}{(\ln 2)^2} \int_{\ln 2}^{2\ln 2} \int_{\ln 2}^{2\ln 2} \frac{1}{xy}\, dy\, dx = \frac{1}{(\ln 2)^2} \int_{\ln 2}^{2\ln 2} \frac{1}{x} \ln 2\, dx = 1$$

55.
$$\iint_R f(x)g(y)\, dx\, dy = \int_c^d \int_a^b f(x)g(y)\, dx\, dy = \int_c^d \left(\int_a^b f(x)g(y)\, dx \right) dy$$

$$= \int_c^d g(y) \left(\int_a^b f(x)\, dx \right) dy = \left(\int_a^b f(x)\, dx \right) \left(\int_c^d g(y)\, dy \right)$$

57. We have $R: -a \le x \le a,\quad c \le y \le d$.   Set $f(x,y) = g_y(x)$.   For each fixed $y \in [c,d]$, g_y is an odd function. Thus

$$\int_{-a}^a g_y(x)\, dx = 0. \qquad (5.8.8)$$

Therefore

$$\iint_R f(x,y)\,dxdy = \int_c^d \int_{-a}^a f(x,y)\,dx\,dy$$

$$= \int_c^d \int_{-a}^a g_y(x)\,dx\,dy = \int_c^d 0\,dy = 0.$$

59. Note that $\quad \Omega = \{(x,y): 0 \le x \le y, \quad 0 \le y \le 1\}.$

Set $\quad \Omega' = \{(x,y): 0 \le y \le x, \quad 0 \le x \le 1\}.$

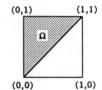

$$\iint_\Omega f(x)f(y)\,dxdy = \int_0^1 \int_0^y f(x)f(y)\,dx\,dy$$

$$= \int_0^1 \int_0^x f(y)f(x)\,dy\,dx$$

x and y are dummy variables

$$= \int_0^1 \int_0^x f(x)f(y)\,dy\,dx = \iint_{\Omega'} f(x)f(y)\,dxdy.$$

Note that Ω and Ω' don't overlap and their union is the unit square

$$R = \{(x,y): 0 \le x \le 1, \quad 0 \le y \le 1\}.$$

If $\quad \displaystyle\int_0^1 f(x)\,dx = 0, \quad$ then

$$0 = \left(\int_0^1 f(x)\,dx\right)\left(\int_0^1 f(y)\,dy\right) = \iint_R f(x)f(y)\,dxdy$$

by Exercise 55

$$= \iint_\Omega f(x)f(y)\,dxdy + \iint_{\Omega'} f(x)f(y)\,dxdy$$

$$= 2\iint_\Omega f(x)f(y)\,dxdy$$

and therefore $\displaystyle\iint_\Omega f(x)f(y)\,dxdy = 0.$

61. Let M be the maximum value of $|f(x,y)|$ on Ω.

$$\int_{\phi_1(x+h)}^{\phi_2(x+h)} = \int_{\phi_1(x+h)}^{\phi_1(x)} + \int_{\phi_1(x)}^{\phi_2(x)} + \int_{\phi_2(x)}^{\phi_2(x+h)}$$

$$|F(x+h) - F(x)| = \left| \int_{\phi_1(x+h)}^{\phi_2(x+h)} f(x,y)\,dy - \int_{\phi_1(x)}^{\phi_2(x)} f(x,y)\,dy \right|$$

$$= \left| \int_{\phi_1(x+h)}^{\phi_1(x)} f(x,y)\,dy + \int_{\phi_2(x)}^{\phi_2(x+h)} f(x,y)\,dy \right|$$

$$\leq \left| \int_{\phi_1(x+h)}^{\phi_1(x)} f(x,y)\,dy \right| + \left| \int_{\phi_2(x)}^{\phi_2(x+h)} f(x,y)\,dy \right|$$

$$\leq \left| \phi_1(x) - \phi_1(x+h) \right| M + \left| \phi_2(x+h) - \phi_2(x) \right| M.$$

The expression on the right tends to 0 as h tends to 0 since ϕ_1 and ϕ_2 are continuous.

SECTION 16.5

1. $\displaystyle \int_0^{\pi/2} \int_0^{\sin\theta} r\cos\theta\,dr\,d\theta = \int_0^{\pi/2} \frac{1}{2}\sin^2\theta\cos\theta\,d\theta = \left[\frac{1}{6}\sin^3\theta\right]_0^{\pi/2} = \frac{1}{6}$

3. $\displaystyle \int_0^{\pi/2} \int_0^{3\sin\theta} r^2\,dr\,d\theta = \int_0^{\pi/2} 9\sin^3\theta\,d\theta = 9\int_0^{\pi/2}(1-\cos^2\theta)\sin\theta\,d\theta = 9\left[-\cos\theta + \frac{1}{3}\cos^3\theta\right]_0^{\pi/2} = 6$

5. (a) $\Gamma: 0 \leq \theta \leq 2\pi, \quad 0 \leq r \leq 1$

$$\iint_\Gamma (\cos r^2)r\,dr\,d\theta = \int_0^{2\pi} \int_0^1 (\cos r^2)r\,dr\,d\theta$$

$$= 2\pi \int_0^1 r\cos r^2\,dr = \pi \sin 1 \cong 0.84\,\pi$$

(b) $\Gamma: 0 \leq \theta \leq 2\pi, \quad 1 \leq r \leq 2$

$$\iint_\Gamma (\cos r^2)r\,dr\,d\theta = \int_0^{2\pi} \int_1^2 (\cos r^2)r\,dr\,d\theta$$

$$= 2\pi \int_1^2 r\cos r^2\,dr = \pi(\sin 2 - \sin 1) \cong 0.07\pi$$

7. (a) $\Gamma: 0 \leq \theta \leq \pi/2, \quad 0 \leq r \leq 1$

$$\iint_{\Gamma} (r\cos\theta + r\sin\theta)r\, drd\theta = \int_0^{\pi/2}\int_0^1 r^2(\cos\theta + \sin\theta)\, dr\, d\theta$$

$$= \left(\int_0^{\pi/2}(\cos\theta + \sin\theta)\, d\theta\right)\left(\int_0^1 r^2\, dr\right) = 2\left(\frac{1}{3}\right) = \frac{2}{3}$$

(b) $\Gamma:\ 0 \le \theta \le \pi/2,\quad 1 \le r \le 2$

$$\iint_{\Gamma} (r\cos\theta + r\sin\theta)r\, drd\theta = \int_0^{\pi/2}\int_1^2 r^2(\cos\theta + \sin\theta)\, dr\, d\theta$$

$$= \left(\int_0^{\pi/2}(\cos\theta + \sin\theta)\, d\theta\right)\left(\int_1^2 r^2\, dr\right) = 2\left(\frac{7}{3}\right) = \frac{14}{3}$$

9. $\displaystyle\int_{-\pi/2}^{\pi/2}\int_0^1 r^2\, dr\, d\theta = \frac{1}{3}\pi$ **11.** $\displaystyle\int_0^{\pi/3}\int_0^1 r^4\, dr\, d\theta = \frac{1}{15}\pi$

13. $\displaystyle\int_0^1\int_0^{\sqrt{1-x^2}} \sin(x^2+y^2)\, dx\, dy = \int_0^{\pi/2}\int_0^1 \sin(r^2)\, r\, dr\, d\theta = \int_0^{\pi/2}\frac{1}{2}(1-\cos 1)\, d\theta = \frac{\pi}{4}(1-\cos 1)$

15. $\displaystyle\int_0^2\int_0^{\sqrt{2x-x^2}} x\, dx\, dy = \int_0^{\pi/2}\int_0^{2\cos\theta} r\cos\theta\, r\, dr\, d\theta = \frac{8}{3}\int_0^{\pi/2}\cos^4\theta\, d\theta = \frac{8}{3}\cdot\frac{3}{4}\cdot\frac{1}{2}\cdot\frac{\pi}{2} = \frac{\pi}{2}$

(See Exercise 46, Section 8.3)

17. $\displaystyle A = \int_0^{\pi/3}\int_0^{3\sin 3\theta} r\, dr\, d\theta = \frac{9}{2}\int_0^{\pi/3}\sin^2 3\theta\, d\theta = \frac{9}{4}\int_0^{\pi/3}(1-6\cos\theta)\, d\theta = \frac{3\pi}{4}$

19. First we find the points of intersection:

$r = 4\cos\theta = 2 \implies \cos\theta = \frac{1}{2}$

$\implies \theta = \pm\dfrac{\pi}{3}.$

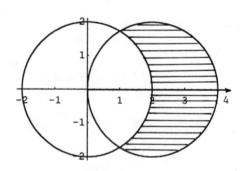

$$A = \int_{-\pi/3}^{\pi/3}\int_2^{4\cos\theta} r\, dr\, d\theta = \int_{-\pi/3}^{\pi/3}(8\cos^2\theta - 2)\, d\theta = \int_{-\pi/3}^{\pi/3}(2 + 4\cos 2\theta)\, d\theta = \frac{4\pi}{3} + 2\sqrt{3}$$

21. $\displaystyle A = 4\int_0^{\pi/4}\int_0^{2\sqrt{\cos 2\theta}} r\, dr\, d\theta = 8\int_0^{\pi/4}\cos 2\theta\, d\theta = 4$

23.
$$\int_0^{2\pi} \int_0^b \left(r^2 \sin\theta + br\right) dr\, d\theta = \int_0^{2\pi} \left[\frac{1}{3}r^3 \sin\theta + \frac{b}{2}r^2\right]_0^b d\theta$$

$$= b^3 \int_0^{2\pi} \left(\frac{1}{3}\sin\theta + \frac{1}{2}\right) d\theta = b^3 \pi$$

25.
$$8\int_0^{\pi/2} \int_0^2 \frac{r}{2}\sqrt{12 - 3r^2}\, dr\, d\theta = 8\int_0^{\pi/2} \left[-\frac{1}{18}\left(12 - 3r^2\right)^{3/2}\right]_0^2 d\theta$$

$$= 8\int_0^{\pi/2} \frac{4}{3}\sqrt{3}\, d\theta = \frac{16}{3}\sqrt{3}\,\pi$$

27.
$$\int_0^{2\pi} \int_0^1 r\sqrt{4 - r^2}\, dr\, d\theta = \int_0^{2\pi} \left[-\frac{1}{3}\left(4 - r^2\right)^{3/2}\right]_0^1 d\theta$$

$$= \int_0^{2\pi} \left(\frac{8}{3} - \sqrt{3}\right) d\theta = \frac{2}{3}(8 - 3\sqrt{3})\pi$$

29.
$$\int_{-\pi/2}^{\pi/2} \int_0^{2\cos\theta} 2r^2 \cos\theta\, dr\, d\theta = \int_{-\pi/2}^{\pi/2} \left[\frac{2}{3}r^3 \cos\theta\right]_0^{2\cos\theta} d\theta$$

$$= \int_{-\pi/2}^{\pi/2} \frac{16}{3}\cos^4\theta\, d\theta = \frac{32}{3}\int_0^{\pi/2} \cos^4\theta\, d\theta = \frac{32}{3}\left(\frac{3}{16}\pi\right) = 2\pi$$

Ex. 46, Sect. 8.3

31.
$$\frac{b}{a}\int_0^{\pi} \int_0^{a\sin\theta} r\sqrt{a^2 - r^2}\, dr\, d\theta = \frac{b}{a}\int_0^{\pi} \left[-\frac{1}{3}\left(a^2 - r^2\right)^{3/2}\right]_0^{a\sin\theta} d\theta$$

$$= \frac{1}{3}a^2 b\int_0^{\pi} \left(1 - \cos^3\theta\right) d\theta = \frac{1}{3}\pi a^2 b$$

SECTION 16.6

1. $$M = \int_{-1}^1 \int_0^1 x^2\, dy\, dx = \frac{2}{3}$$

$$x_M M = \int_{-1}^1 \int_0^1 x^3\, dy\, dx = 0 \quad \Longrightarrow \quad x_M = 0$$

$$y_M M = \int_{-1}^1 \int_0^1 x^2 y\, dy\, dx = \int_{-1}^1 \frac{1}{2}x^2\, dx = \frac{1}{3} \quad \Longrightarrow \quad y_M = \frac{1/3}{1/2} = \frac{1}{2}$$

3. $M = \int_0^1 \int_{x^2}^1 xy\,dy\,dx = \frac{1}{2}\int_0^1 (x - x^5)\,dx = \frac{1}{6}$

$x_M M = \int_0^1 \int_{x^2}^1 x^2 y\,dy\,dx \frac{1}{2}\int_0^1 (x^2 - x^6)\,dx = \frac{2}{21} \implies x_M = \frac{2/21}{1/6} = \frac{12}{21}$

$y_M M = \int_0^1 \int_{x^2}^1 xy^2\,dy\,dx = \frac{1}{3}\int_0^1 (x - x^7)\,dx = \frac{1}{8} \implies y_M = \frac{1/8}{1/6} = \frac{3}{4}$

5. $M = \int_0^8 \int_0^{x^{1/3}} y^2\,dy\,dx = \frac{1}{3}\int_0^8 x\,dx = \frac{32}{3}$

$x_M M = \int_0^8 \int_0^{x^{1/3}} xy^2\,dy\,dx \frac{1}{3}\int_0^8 x^2\,dx = \frac{512}{9} \implies x_M = \frac{512/9}{32/3} = \frac{16}{3}$

$y_M M = \int_0^8 \int_0^{x^{1/3}} y^3\,dy\,dx = \frac{1}{4}\int_0^8 x^{4/3}\,dx = \frac{96}{7} \implies y_M = \frac{96/7}{32/3} = \frac{9}{7}$

7. $M = \int_0^1 \int_{2x}^{3x} xy\,dy\,dx = \frac{5}{2}\int_0^1 x^3\,dx = \frac{5}{8}$

$x_M M = \int_0^1 \int_{2x}^{3x} x^2 y\,dy\,dx = \frac{5}{2}\int_0^1 x^4\,dx = \frac{1}{2} \implies x_M = \frac{1/2}{5/8} = \frac{4}{5}$

$y_M M = \int_0^1 \int_{2x}^{3x} xy^2\,dy\,dx = \frac{19}{3}\int_0^1 x^4\,dx = \frac{19}{15} \implies y_M = \frac{19/15}{5/8} = \frac{152}{75}$

9. $M = \int_0^{2\pi} \int_0^{1+\cos\theta} r^2\,dr\,d\theta = \frac{1}{3}\int_0^{2\pi} (1 + 3\cos\theta + 3\cos^2\theta + \cos^3\theta)\,d\theta = \frac{5\pi}{3}$

$x_M M = \int_0^{2\pi} \int_0^{1+\cos\theta} r^3\cos\theta\,dr\,d\theta = \frac{1}{4}\int_0^{2\pi} (1+\cos\theta)^4\cos\theta\,d\theta$

$= \frac{1}{4}\int_0^{2\pi} \left[\cos\theta + 4\cos^2\theta + 6\cos^3\theta + 4\cos^4\theta + \cos^5\theta\right]\,d\theta$

$= \frac{7\pi}{4}$

Therefore, $x_M = \frac{7\pi/4}{5\pi/3} = \frac{21}{20}$.

$y_M M = \int_0^{2\pi} \int_0^{1+\cos\theta} r^3\sin\theta\,dr\,d\theta = \frac{1}{4}\int_0^{2\pi} (1+\cos\theta)^4\sin\theta\,d\theta = \frac{1}{4}\left[\frac{1}{5}(1+\cos\theta)^5\right]_0^{2\pi} = 0$

Therefore, $y_M = 0$.

11. $\Omega: -L/2 \le x \le L/2, \quad -W/2 \le y \le W/2$

$$I_x = \iint_\Omega \frac{M}{LW} y^2\, dxdy = \frac{4M}{LW} \int_0^{W/2} \int_0^{L/2} y^2\, dx\, dy = \frac{1}{12} MW^2$$

symmetry

$$I_y = \iint_\Omega \frac{M}{LW} x^2\, dxdy = \frac{1}{12} ML^2, \quad I_z = \iint_\Omega \frac{M}{LW}(x^2+y^2)\, dxdy = \frac{1}{12} M(L^2+W^2)$$

$$K_x = \sqrt{I_x/M} = W/2\sqrt{3}, \quad K_y = \sqrt{I_y/M} = L/2\sqrt{3}$$

$$K_z = \sqrt{I_z/M} = \sqrt{L^2+W^2}\Big/2\sqrt{3}$$

13.
$$M = \iint_\Omega k\left(x+\frac{L}{2}\right) dxdy = \iint_\Omega \frac{1}{2} kL\, dxdy = \frac{1}{2} kL(\text{ area of } \Omega) = \frac{1}{2} kL^2 W$$

symmetry

$$x_M M = \iint_\Omega x\left[k\left(x+\frac{L}{2}\right)\right] dxdy = \iint_\Omega \left(kx^2 + \frac{1}{2} Lx\right) dxdy$$

$$= \iint_\Omega kx^2\, dxdy = 4k \int_0^{W/2} \int_0^{L/2} x^2\, dx\, dy = \frac{1}{12} kWL^3$$

symmetry symmetry

$$= \frac{1}{6}\left(\frac{1}{2} kL^2 W\right) L = \frac{1}{6} ML; \quad x_M = \frac{1}{6} L$$

$$y_M M = \iint_\Omega y\left[k\left(x+\frac{L}{2}\right)\right] dxdy = 0; \quad y_M = 0$$

by symmetry

15.
$$I_x = \iint_\Omega \frac{4M}{\pi R^2} y^2\, dxdy = \frac{4M}{\pi R^2} \int_0^{\pi/2} \int_0^R r^3 \sin^2\theta\, dr\, d\theta$$

$$= \frac{4M}{\pi R^2}\left(\int_0^{\pi/2} \sin^2\theta\, d\theta\right)\left(\int_0^R r^3\, dr\right) = \frac{4M}{\pi R^2}\left(\frac{\pi}{4}\right)\left(\frac{1}{4} R^4\right) = \frac{1}{4} MR^2$$

$$I_y = \frac{1}{4} MR^2, \quad I_z = \frac{1}{2} MR^2$$

$$K_x = K_y = \frac{1}{2} R, \quad K_z = R/\sqrt{2}$$

17. I_M, the moment of inertia about the vertical line through the center of mass, is

$$\iint_\Omega \frac{M}{\pi R^2}\left(x^2 + y^2\right) dx\, dy$$

where Ω is the disc of radius R centered at the origin. Therefore

$$I_M = \frac{M}{\pi R^2}\int_0^{2\pi}\int_0^R r^3\, dr\, d\theta = \frac{1}{2}MR^2.$$

We need $I_0 = \frac{1}{2}MR^2 + d^2 M$ where d is the distance from the center of the disc to the origin. Solving this equation for d, we have $d = \sqrt{I_0 - \frac{1}{2}MR^2}\Big/\sqrt{M}$.

19. $\Omega : 0 \le x \le a, \quad 0 \le y \le b$

$$I_x = \iint_\Omega \frac{4M}{\pi ab} y^2\, dx\, dy = \frac{4M}{\pi ab}\int_0^a \int_0^{\frac{b}{a}\sqrt{a^2-x^2}} y^2\, dy\, dx = \frac{1}{4}Mb^2$$

$$I_y = \iint_\Omega \frac{4M}{\pi ab} x^2\, dx\, dy = \frac{4M}{\pi ab}\int_0^a \int_0^{\frac{b}{a}\sqrt{a^2-x^2}} x^2\, dy\, dx = \frac{1}{4}Ma^2$$

$$I_z = \frac{1}{4}M\left(a^2 + b^2\right)$$

21. $I_x = \int_0^1 \int_{x^2}^1 xy^3\, dy\, dx = \frac{1}{4}\int_0^1 (x - x^9)\, dx = \frac{1}{10}$

$$I_y = \int_0^1 \int_{x^2}^1 x^3 y\, dy\, dx\, \frac{1}{2}\int_0^1 (x^3 - x^7)\, dx = \frac{1}{16}$$

$$I_z = \int_0^1 \int_{x^2}^1 xy(x^2 + y^2)\, dy\, dx = I_x + I_y = \frac{13}{80}$$

23. $I_x = \int_0^{2\pi}\int_0^{1+\cos\theta} r^4 \sin^2\theta\, dr\, d\theta = \frac{1}{5}\int_0^{2\pi}(1+\cos\theta)^5 \sin^2\theta\, d\theta = \frac{33\pi}{40}$

$$I_y = \int_0^{2\pi}\int_0^{1+\cos\theta} r^4 \cos^2\theta\, dr\, d\theta = \frac{1}{5}\int_0^{2\pi}(1+\cos\theta)^5 \cos^2\theta\, d\theta = \frac{93\pi}{40}$$

$$I_z = \int_0^{2\pi}\int_0^{1+\cos\theta} r^4\, dr\, d\theta = I_x + I_y = \frac{63\pi}{20}$$

25. $\Omega : r_1^2 \le x^2 + y^2 \le r_2^2, \quad A = \pi\left(r_2^2 - r_1^2\right)$

(a) Place the diameter on the x-axis.

$$I_x = \iint_\Omega \frac{M}{A} y^2\, dx\, dy = \frac{M}{A}\int_0^{2\pi}\int_{r_1}^{r_2}\left(r^2 \sin^2\theta\right) r\, dr\, d\theta = \frac{1}{4}M\left(r_2^2 + r_1^2\right)$$

(b) $\frac{1}{4}M\left(r_2^2 + r_1^2\right) + Mr_1^2 = \frac{1}{4}M\left(r_2^2 + 5r_1^2\right)$ (parallel axis theorem)

(c) $\frac{1}{4}M\left(r_2{}^2 + r_1{}^2\right) + Mr_2{}^2 = \frac{1}{4}M\left(5r_2{}^2 + r_1{}^2\right)$

27. $\Omega: \;\; r_1{}^2 \le x^2 + y^2 \le r_2{}^2, \quad A = \pi\left(r_2{}^2 - r_1{}^2\right)$

$$I = \iint\limits_{\Omega} \frac{M}{A}\left(x^2 + y^2\right) dxdy = \frac{M}{A}\int_0^{2\pi}\int_{r_1}^{r_2} r^3\, dr\, d\theta = \frac{1}{2}M(r_2{}^2 + r_1{}^2)$$

29. $M = \iint\limits_{\Omega} k\left(R - \sqrt{x^2 + y^2}\right) dxdy = k\int_0^{\pi}\int_0^R \left(Rr - r^2\right) dr\, d\theta = \frac{1}{6}k\pi R^3$

$x_M = 0$ by symmetry

$$y_M M = \iint\limits_{\Omega} y\left[k\left(R - \sqrt{x^2 + y^2}\right)\right] dxdy = k\int_0^{\pi}\int_0^R \left(Rr^2 - r^3\right)\sin\theta\, dr\, d\theta = \frac{1}{6}kR^4$$

$y_M = R/\pi$

31.   Place P at the origin.

$$M = \iint\limits_{\Omega} k\sqrt{x^2 + y^2}\; dxdy$$

$$= k\int_0^{\pi}\int_0^{2R\sin\theta} r^2\, dr\, d\theta = \frac{32}{9}kR^3$$

$x_M = 0$ by symmetry

$$y_M M = \iint\limits_{\Omega} y\left(k\sqrt{x^2 + y^2}\right) dxdy = k\int_0^{\pi}\int_0^{2R\sin\theta} r^3\sin\theta\, dr\, d\theta = \frac{64}{15}kR^4$$

$y_M = 6R/5$

Answer: the center of mass lies on the diameter through P at a distance $6R/5$ from P.

33.   Suppose Ω, a basic region of area A, is broken up into n basic regions $\Omega_1, \cdots, \Omega_n$ with areas $A_1, \cdots, A_n$. Then

$$\overline{x}A = \iint\limits_{\Omega} x\, dxdy = \sum_{i=1}^n \left(\iint\limits_{\Omega_i} x\, dxdy\right) = \sum_{i=1}^n \overline{x}_i\, A_i = \overline{x}_1 A_1 + \cdots + \overline{x}_n A_n.$$

The second formula can be derived in a similar manner.

SECTION 16.7

1. They are equal; they both give the volume of T.

3. $\displaystyle\iiint_\Pi \alpha \, dx \, dy \, dz = \alpha \iiint_\Pi dx \, dy \, dz = \alpha \left[\text{volume (II)}\right] = \alpha(a_2 - a_1)(b_2 - b_1)(c_2 - c_1)$

5. Let $P_1 = \{x_0, \cdots, x_m\}$, $\quad P_2 = \{y_0, \cdots, y_n\}$, $\quad P_3 = \{z_0, \cdots, z_q\}$ be partitions of $[0,a]$, $[0,b]$, $[0,c]$ respectively and let $P = P_1 \times P_2 \times P_3$. Note that

$$x_{i-1}y_{j-1} \leq \left(\frac{x_i + x_{i-1}}{2}\right)\left(\frac{y_j + y_{j-1}}{2}\right) \leq x_i y_j$$

and therefore

$$x_{i-1}y_{j-1}\,\Delta x_i\,\Delta y_j\,\Delta z_k \leq \tfrac{1}{4}\left(x_i{}^2 - x_{i-1}^2\right)\left(y_j{}^2 - y_{j-1}^2\right)\Delta z_k \leq x_i y_j\,\Delta x_i\,\Delta y_j\,\Delta z_k.$$

It follows that

$$L_f(P) \leq \frac{1}{4}\sum_{i=1}^{m}\sum_{j=1}^{n}\sum_{k=1}^{q}\left(x_i{}^2 - x_{i-1}^2\right)\left(y_j{}^2 - y_{j-1}^2\right)\Delta z_k \leq U_f(P).$$

The middle term can be written

$$\frac{1}{4}\left(\sum_{i=1}^{m}x_i{}^2 - x_{i-1}^2\right)\left(\sum_{j=1}^{n}y_j{}^2 - y_{j-1}^2\right)\left(\sum_{k=1}^{q}\Delta z_k\right) = \frac{1}{4}\,a^2 b^2 c.$$

7. $\bar{x}_1 = a$, $\quad \bar{y}_1 = b$, $\quad \bar{z}_1 = c$; $\quad\quad \bar{x}_0 = A$, $\quad \bar{y}_0 = B$, $\quad \bar{z}_0 = C$

$$\bar{x}_1 V_1 + \bar{x} V = \bar{x}_0 V_0 \quad\Longrightarrow\quad a^2 bc + (ABC - abc)\bar{x} = A^2 BC$$

$$\Longrightarrow\quad \bar{x} = \frac{A^2 BC - a^2 bc}{ABC - abc}$$

similarly

$$\bar{y} = \frac{AB^2 C - ab^2 c}{ABC - abc}, \quad \bar{z} = \frac{ABC^2 - abc^2}{ABC - abc}$$

9. $\displaystyle M = \iiint_\Pi Kz \, dx\,dy\,dz$

Let $P_1 = \{x_0, \cdots, x_m\}$, $\quad P_2 = \{y_0, \cdots, y_n\}$, $\quad P_3 = \{z_0, \cdots, z_q\}$ be partitions of $[0,a]$ $\quad$ and let $P = P_1 \times P_2 \times P_3$. $\quad$ Note that

$$z_{k-1} \leq \tfrac{1}{2}\left(z_k + z_{k-1}\right) \leq z_k$$

and therefore

$$Kz_{k-1}\,\Delta x_i\,\Delta y_j\,\Delta z_k \leq \tfrac{1}{2}K\,\Delta x_i\,\Delta y_j\left(z_k{}^2 - z_{k-1}^2\right) \leq Kz_k\,\Delta x_i\,\Delta y_j\,\Delta z_k.$$

It follows that

$$L_f(P) \le \frac{1}{2} K \sum_{i=1}^{m} \sum_{j=1}^{n} \sum_{k=1}^{q} \Delta x_i \, \Delta y_j \, (z_k{}^2 - z_{k-1}^2) \le U_f(P).$$

The middle term can be written

$$\frac{1}{2} K \left(\sum_{i=1}^{m} \Delta x_i \right) \left(\sum_{j=1}^{n} \Delta y_j \right) \left(\sum_{k=1}^{q} z_k{}^2 - z_{k-1}^2 \right) = \frac{1}{2} K(a)(a)(a^2) = \frac{1}{2} K a^4.$$

$M = \frac{1}{2} K a^4$ where K is the constant of proportionality for the density function.

11.
$$I_z = \iiint_{\Pi} Kz \left(x^2 + y^2 \right) dx dy dz$$

$$= \underbrace{\iiint_{\Pi} Kx^2 z \, dx dy dz}_{I_1} + \underbrace{\iiint_{\Pi} Ky^2 z \, dx dy dz}_{I_2} \, .$$

We will calculate I_1 using the partitions we used in doing Exercise 9.   Note that

$$x_{i-1}^2 z_{k-1} \le \left(\frac{x_i{}^2 + x_i x_{i-1} + x_{i-1}^2}{3} \right) \left(\frac{z_k + z_{k-1}}{2} \right) \le x_i{}^2 z_k$$

and therefore

$$K x_{i-1}^2 z_{k-1} \Delta x_i \, \Delta y_j \, \Delta z_k \le \frac{1}{6} K \left(x_i{}^3 - x_{i-1}^3 \right) \Delta y_j \, (z_k{}^2 - z_{k-1}^2) \le K x_i{}^2 z_k{}^2 \, \Delta x_i \, \Delta y_j \, \Delta z_k.$$

It follows that

$$L_f(P) \le \frac{1}{6} K \sum_{i=1}^{m} \sum_{j=1}^{n} \sum_{k=1}^{q} \left(x_i{}^3 - x_{i-1}^3 \right) \Delta y_j \, (z_k{}^2 - z_{k-1}^2) \le U_f(P).$$

The middle term can be written

$$\frac{1}{6} K \left(\sum_{i=1}^{m} x_i{}^3 - x_{i-1}^3 \right) \left(\sum_{j=1}^{n} \Delta y_j \right) \left(\sum_{k=1}^{q} z_k{}^2 - z_{k-1}^2 \right) = \frac{1}{6} K a^3 (a)(a^2) = \frac{1}{6} K a^6.$$

Similarly $I_2 = \frac{1}{6} K a^6$ and therefore

by Exercise 9

$$I_z = \frac{1}{3} K a^6 = \frac{2}{3} \left(\frac{1}{2} K a^4 \right) a^2 = \frac{2}{3} M a^2.$$

SECTION 16.8

1. $\displaystyle \int_0^a \int_0^b \int_0^c dx \, dy \, dz = \int_0^a \int_0^b c \, dy \, dz = \int_0^a bc \, dz = abc$

3. $\displaystyle \int_0^1 \int_1^{2y} \int_0^x (x + 2z) \, dz \, dx \, dy = \int_0^1 \int_1^{2y} \left[xz + z^2 \right]_0^x dx \, dy = \int_0^1 \int_1^{2y} 2x^2 \, dx \, dy$

$$= \int_0^1 \left[\frac{2}{3} x^3 \right]_1^{2y} dy = \int_0^1 \left(\frac{16}{3} y^3 - \frac{2}{3} \right) dy = \frac{2}{3}$$

5. $\displaystyle\int_0^2 \int_{-1}^1 \int_0^3 (z - xy)\, dz\, dy\, dx = \int_0^2 \int_{-1}^1 \left[\frac{1}{2}z^2 - xyz\right]_1^3 dy\, dx$

$\displaystyle = \int_0^2 \int_{-1}^1 (4 - 2xy)\, dy\, dx = \int_0^2 \left[2y - xy^2\right]_{-1}^1 dx = \int_0^2 8\, dy = 16$

7. $\displaystyle\int_0^{\pi/2} \int_0^1 \int_0^{\sqrt{1-x^2}} (x \cos z\, dy\, dx\, dz = \int_0^{\pi/2} \int_0^1 [xy \cos z]_0^{\sqrt{1-x^2}}\, dx\, dz$

$\displaystyle = \int_0^{\pi/2} \int_0^1 x\sqrt{1-x^2} \cos z\, dx\, dz = \int_0^{\pi/2} \left[-\frac{1}{3}(1-x^2)^{3/2} \cos z\right]_0^1 dz = \frac{1}{3}\int_0^{\pi/2} \cos z\, dz = \frac{1}{3}$

9. $\displaystyle\int_1^2 \int_y^{y^2} \int_0^{\ln x} ye^z\, dz\, dx\, dy = \int_1^2 \int_y^{y^2} [ye^z]_0^{\ln x}\, dx\, dy$

$\displaystyle = \int_1^2 \int_y^{y^2} y(x-1)\, dx\, dy = \int_1^2 \left[\frac{1}{2}x^2 y - xy\right]_y^{y^2} dy = \int_1^2 \left(\frac{1}{2}y^5 - \frac{3}{2}y^3 + y^2\right) dy = \frac{47}{24}$

11. $\displaystyle\iiint_\Pi f(x)g(y)h(z)\, dx\,dy\,dz = \int_{c_1}^{c_2} \left[\int_{b_1}^{b_2} \left(\int_{a_1}^{a_2} f(x)g(y)h(z)\, dx\right) dy\right] dz$

$\displaystyle = \int_{c_1}^{c_2} \left[\int_{b_1}^{b_2} g(y)h(z) \left(\int_{a_1}^{a_2} f(x)\, dx\right) dy\right] dz$

$\displaystyle = \int_{c_1}^{c_2} \left[h(z) \left(\int_{a_1}^{a_2} f(x)\, dx\right) \left(\int_{b_1}^{b_2} g(y)\, dy\right) dz\right]$

$\displaystyle = \left(\int_{a_1}^{a_2} f(x)\, dx\right) \left(\int_{b_1}^{b_2} g(y)\, dy\right) \left(\int_{c_1}^{c_2} h(z)\, dz\right)$

13. $\displaystyle\left(\int_0^1 x^2\, dx\right) \left(\int_0^2 y^2\, dy\right) \left(\int_0^3 z^2\, dz\right) = \left(\frac{1}{3}\right)\left(\frac{8}{3}\right)\left(\frac{27}{3}\right) = 8$

15. $\displaystyle x_M M = \iiint_\Pi kx^2 yz\, dx\,dy\,dz = k\left(\int_0^a x^2\, dx\right)\left(\int_0^b y\, dy\right)\left(\int_0^c z\, dz\right)$

$\displaystyle = k\left(\tfrac{1}{3}a^3\right)\left(\tfrac{1}{2}b^2\right)\left(\tfrac{1}{2}c^2\right) = \tfrac{1}{12}ka^3 b^2 c^2.$

By Exercise 14, $M = \frac{1}{8}ka^2 b^2 c^2$. Therefore $\bar{x} = \frac{2}{3}a$. Similarly, $\bar{y} = \frac{2}{3}b$ and $\bar{z} = \frac{2}{3}c$.

17.

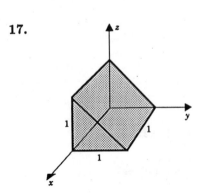

19.   center of mass is the centroid

$$\bar{x} = \tfrac{1}{2} \quad \text{by symmetry}$$

$$\bar{y}V = \iiint_T y\, dx\, dy\, dz = \int_0^1 \int_0^1 \int_0^{1-y} y\, dz\, dy\, dx = \int_0^1 \int_0^1 (y - y^2)\, dy\, dx$$

$$= \int_0^1 \left[\frac{1}{2} y^2 - \frac{1}{3} y^3 \right]_0^1 dx = \int_0^1 \frac{1}{6}\, dx = \frac{1}{6}$$

$$\bar{z}V = \iiint_T z\, dx\, dy\, dz = \int_0^1 \int_0^1 \int_0^{1-y} z\, dz\, dy\, dx = \int_0^1 \int_0^1 \frac{1}{2}(1-y)^2\, dy\, dx$$

$$= \frac{1}{2} \int_0^1 \int_0^1 (1 - 2y + y^2)\, dy\, dx = \frac{1}{2} \int_0^1 \left[y - y^2 \frac{1}{3} y^3 \right]_0^1 dx = \frac{1}{2} \int_0^1 \frac{1}{3}\, dx = \frac{1}{6}$$

$V = \tfrac{1}{2}$ (by Exercise 18); $\bar{y} = \tfrac{1}{3}$, $\bar{z} = \tfrac{1}{3}$

21. $\displaystyle \int_{-r}^{r} \int_{-\phi(x)}^{\phi(x)} \int_{-\psi(x,y)}^{\psi(x,y)} k\left(r - \sqrt{x^2 + y^2 + z^2} \right) dz\, dy\, dx$ with $\phi(x) = \sqrt{r^2 - x^2}$,

$\psi(x,y) = \sqrt{r^2 - (x^2 + y^2)}$, k the constant of proportionality

23. $\displaystyle \int_0^1 \int_{-\sqrt{x-x^2}}^{\sqrt{x-x^2}} \int_{-2x-3y-10}^{1-y^2} dz\, dy\, dx$

25. $\displaystyle \int_{-1}^{1} \int_{-2\sqrt{2-2x^2}}^{2\sqrt{2-2x^2}} \int_{3x^2+y^2/4}^{4-x^2-y^2/4} k\left(z - 3x^2 - \frac{1}{4} y^2 \right) dz\, dy\, dx$

27. $\displaystyle \iiint_T (x^2 z + y)\, dx\, dy\, dz = \int_0^2 \int_1^3 \int_0^1 (x^2 z + y)\, dx\, dy\, dz = \int_0^2 \int_1^3 \left[\frac{1}{3} x^3 z + xy \right]_0^1 dy\, dz$

$$= \int_0^2 \int_1^3 \left(\frac{1}{3} z + y \right) dy\, dz = \int_0^2 \left[\frac{1}{3} zy + \frac{1}{2} y^2 \right]_1^3 dz = \int_0^2 \left(\frac{2}{3} z + 4 \right) dz = \frac{28}{3}$$

29. $\displaystyle \iiint_T x^2 y^2 z^2\, dx\, dy\, dz = \int_{-1}^{0} \int_0^{y+1} \int_0^1 x^2 y^2 z^2\, dx\, dz\, dy + \int_0^1 \int_0^{1-y} \int_0^1 x^2 y^2 z^2\, dx\, dz\, dy$

$$= \int_{-1}^{0} \int_0^{y+1} \left[\frac{1}{3} x^3 y^2 z^2 \right]_0^1 dz\, dy + \int_0^1 \int_0^{1-y} \left[\frac{1}{3} x^3 y^2 z^2 \right]_0^1 dz\, dy$$

$$= \frac{1}{3} \int_{-1}^{0} \int_0^{y+1} y^2 z^2\, dz\, dy + \frac{1}{3} \int_0^1 \int_0^{1-y} [y^2 z^2]_0^1 dz\, dy$$

$$= \frac{1}{3} \int_{-1}^{0} \left[\frac{1}{3} y^2 z^3 \right]_0^{y+1} dy + \frac{1}{3} \int_0^1 \left[\frac{1}{3} y^2 z^3 \right]_0^{1-y} dy$$

$$= \frac{1}{9} \int_{-1}^{0} \left(y^5 + 3y^4 + 3y^3 + y^2 \right) dy + \frac{1}{9} \int_{0}^{1} \left(y^2 - 3y^3 + 3y^4 - y^5 \right) dy = \frac{1}{270}$$

31. $$\iiint_T y^2 \, dx \, dy \, dz = \int_0^3 \int_0^{2-2x/3} \int_0^{6-2x-3y} y^2 \, dz \, dy \, dx = \int_0^3 \int_0^{2-2x/3} \left[y^2 z \right]_0^{6-2x-3y} dy \, dx$$

$$= \int_0^3 \int_0^{2-2x/3} \left(6y^2 - 2xy^2 - 3y^3 \right) dy \, dx$$

$$= \int_0^3 \left[2y^3 - \frac{2}{3} xy^3 - \frac{3}{4} y^4 \right]_0^{2-2x/3} dx$$

$$= \frac{1}{4} \int_0^3 \left(2 - \frac{2}{3} x \right) dx = \frac{12}{5}$$

33. $$V = \int_0^2 \int_{x^2}^{x+2} \int_0^x dz \, dy \, dx = \int_0^2 \int_{x^2}^{x+2} x \, dy \, dx = \int_0^2 \left(x^2 + 2x - x^3 \right) dx = \frac{8}{3}$$

$$\overline{x} V = \int_0^2 \int_{x^2}^{x+2} \int_0^x x \, dz \, dy \, dx = \int_0^2 \int_{x^2}^{x+2} x^2 \, dy \, dx = \int_0^2 \left(x^3 + 2x^2 - x^4 \right) dx = \frac{44}{15}$$

$$\overline{y} V = \int_0^2 \int_{x^2}^{x+2} \int_0^x y \, dz \, dy \, dx = \int_0^2 \int_{x^2}^{x+2} xy \, dy \, dx = \int_0^2 \frac{1}{2} \left(x^3 + 4x^2 + 4x - x^5 \right) dx = 6$$

$$\overline{z} V = \int_0^2 \int_{x^2}^{x+2} \int_0^x z \, dz \, dy \, dx = \int_0^2 \int_{x^2}^{x+2} \frac{1}{2} x^2 \, dy \, dx = \int_0^2 \frac{1}{2} \left(x^3 + 2x^2 - x^4 \right) dx = \frac{22}{15}$$

$$\overline{x} = \frac{11}{10}, \quad \overline{y} = \frac{9}{4}, \quad \overline{z} = \frac{11}{20}$$

35. $$V = \int_{-1}^2 \int_0^3 \int_{2-x}^{4-x^2} dz \, dy \, dx = \frac{27}{2}; \quad (\overline{x}, \overline{y}, \overline{z}) = \left(\frac{1}{2}, \frac{3}{2}, \frac{12}{5} \right)$$

37. $$V = \int_0^a \int_0^{\phi(x)} \int_0^{\psi(x,y)} dz \, dy \, dx = \frac{1}{6} abc \quad \text{with} \quad \phi(x) = b \left(1 - \frac{x}{a} \right), \quad \psi(x,y) = c \left(1 - \frac{x}{a} - \frac{y}{b} \right)$$

$$(\overline{x}, \overline{y}, \overline{z}) = \left(\tfrac{1}{4} a, \tfrac{1}{4} b, \tfrac{1}{4} c \right)$$

39. $$\Pi : 0 \leq x \leq a, \quad 0 \leq y \leq b, \quad 0 \leq z \leq c$$

(a) $$I_z = \int_0^a \int_0^b \int_0^c \frac{M}{abc} \left(x^2 + y^2 \right) dz \, dy \, dx = \frac{1}{3} M \left(a^2 + b^2 \right)$$

(b) $$I_M = I_z - d^2 M = \tfrac{1}{3} M \left(a^2 + b^2 \right) - \tfrac{1}{4} \left(a^2 + b^2 \right) M = \tfrac{1}{12} M \left(a^2 + b^2 \right)$$

parallel axis theorem (16.6.7)

(c) $I = I_M + d^2 M = \frac{1}{12} M(a^2 + b^2) + \frac{1}{4} a^2 M = \frac{1}{3} Ma^2 + \frac{1}{12} Mb^2$

└──── parallel axis theorem (16.6.7)

41. $M = \int_0^1 \int_0^1 \int_0^y k\,(x^2 + y^2 + z^2)\,dz\,dy\,dx = \int_0^1 \int_0^1 k\left(x^2 y + y^3 + \frac{1}{3} y^3\right) dy\,dx$

$$= \int_0^1 k\left(\frac{1}{2} x^2 + \frac{1}{3}\right) dx = \frac{1}{2} k$$

$(x_M, y_M, z_M) = \left(\frac{7}{12}, \frac{34}{45}, \frac{37}{90}\right)$

43. (a) 0 by symmetry

(b) $\iiint\limits_T (a_1 x + a_2 y + a_3 z + a_4)\,dx\,dy\,dz = \iiint\limits_T a_4\,dx\,dy\,dz = a_4 \text{ (volume of ball)} = \frac{4}{3}\pi a_4$

by symmetry ──┘

45. $V = 8\int_0^a \int_0^{\sqrt{a^2-x^2}} \int_0^{\sqrt{a^2-x^2-y^2}} dz\,dy\,dx = 8\int_0^a \int_0^{\sqrt{a^2-x^2}} \sqrt{a^2 - x^2 - y^2}\,dy\,dx$

polar coordinates ──┐

$$= 8\int_0^{\pi/2} \int_0^a \sqrt{a^2 - r^2}\,r\,dr\,d\theta$$

$$= -4\int_0^{\pi/2} \left[\frac{2}{3}(a^2 - r^2)^{3/2}\right]_0^a d\theta$$

$$= \frac{8}{3}\int_0^{\pi/2} d\theta = \frac{4}{3}\pi a^3$$

47. $M = \int_{-2}^2 \int_{-\sqrt{4-x^2}/2}^{\sqrt{4-x^2}/2} \int_{x^2+3y^2}^{4-y^2} k|x|\,dz\,dy\,dx = 4\int_0^2 \int_0^{\sqrt{4-x^2}/2} \int_{x^2+3y^2}^{4-y^2} kx\,dz\,dy\,dx$

$$= 4k\int_0^2 \int_0^{\sqrt{4-x^2}/2} \left(4x - x^3 - 4xy^2\right) dy\,dx = \frac{4}{3}k\int_0^2 x\left(4 - x^2\right)^{3/2} dx = \frac{128}{15}k$$

49. $M = \int_{-1}^2 \int_0^3 \int_{2-x}^{4-x^2} k(1+y)\,dz\,dy\,dx = \frac{135}{4}k;\quad (x_M, y_M, z_M) = \left(\frac{1}{2}, \frac{9}{5}, \frac{12}{5}\right)$

51. (a) $V = \int_0^6 \int_{z/2}^3 \int_x^{6-x} dy\,dx\,dz$

(b) $V = \int_0^3 \int_0^{2x} \int_x^{6-x} dy\,dz\,dx$

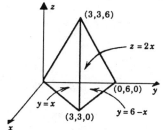

(c) $V = \int_0^6 \int_{z/2}^3 \int_{z/2}^y dx\, dy\, dz + \int_0^6 \int_3^{(12-z)/2} \int_{z/2}^{6-y} dx\, dy\, dz$

53.   (a) $V = \iint\limits_{\Omega_{yz}} 2y\, dy\, dz$ (b) $V = \iint\limits_{\Omega_{yz}} \left(\int_{-y}^y dx \right) dy\, dz$

   (c) $V = \int_0^4 \int_{-\sqrt{4-y}}^{\sqrt{4-y}} \int_{-y}^y dx\, dz\, dy$ (d) $V = \int_{-2}^2 \int_0^{4-z^2} \int_{-y}^y dx\, dy\, dz$

SECTION 16.9

1. $r^2 + z^2 = 9$ **3.** $z = 2r$ **5.** $4r^2 = z^2$

7. $\int_0^\pi \int_0^2 \int_0^{4-r^2} r\, dz\, dr\, d\theta = \int_0^\pi \int_0^2 (4r - r^2)\, dr\, d\theta = \int_0^\pi 4\, d\theta = 4\pi$

9. $\int_0^{\pi/4} \int_0^1 \int_0^{r\cos\theta} r\sec^3\theta\, dz\, dr\, d\theta = \int_0^{\pi/4} \int_0^1 r^2 \sec^2\theta\, dr\, d\theta = \frac{1}{3}\int_0^{\pi/4} \sec^2\theta\, d\theta = \frac{1}{3}$

11.   Set the lower base of the cylinder on the xy-plane so that the axis of the cylinder coincides with the z-axis. Assume that the density varies directly as the distance from the lower base.

$$M = \int_0^{2\pi} \int_0^R \int_0^h kzr\, dz\, dr\, d\theta = \frac{1}{2} k\pi R^2 h^2$$

13.

$$I = I_z = k \int_0^{2\pi} \int_0^R \int_0^h zr^3\, dr\, d\theta\, dz$$

$$= \tfrac{1}{4} k\pi R^4 h^2 = \tfrac{1}{2} \left(\tfrac{1}{2} k\pi R^2 h^2 \right) R^2 = \tfrac{1}{2} MR^2$$

from Exercise 11 ⌐⌐

15.   Inverting the cone and placing the vertex at the origin, we have

$$V = \int_0^h \int_0^{2\pi} \int_0^{(R/h)z} r\, dr\, d\theta\, dz = \frac{1}{3}\pi R^2 h.$$

17. $I = \frac{M}{V} \int_0^h \int_0^{2\pi} \int_0^{(R/h)z} r^3\, dr\, d\theta\, dz = \frac{3}{10} MR^2$

19. $V = \int_0^{2\pi} \int_0^1 \int_0^{1-r^2} r\, dz\, dr\, d\theta = \frac{1}{2}\pi$

21. $M = \int_0^{2\pi} \int_0^1 \int_0^{1-r^2} k(r^2 + z^2)r\, dz\, dr\, d\theta = \frac{1}{4}k\pi$

23. $\int_0^{\pi/2} \int_0^1 \int_0^{\sqrt{4-r^2}} r \, dz \, dr \, d\theta = \int_0^{\pi/2} \int_0^1 r\sqrt{4-r^2} \, dr \, d\theta = \int_0^{\pi/2} \left(\frac{8}{3} - \sqrt{3}\right) d\theta = \frac{1}{6}\left(8 - 3\sqrt{3}\right)\pi$

25. $\int_0^3 \int_0^{\sqrt{9-y^2}} \int_0^{\sqrt{9-x^2y^2}} \frac{1}{\sqrt{x^2+y^2}} \, dz \, dx \, dy = \int_0^{\pi/2} \int_0^3 \int_0^{\sqrt{9-r^2}} \frac{1}{r} \cdot r \, dz \, dr \, d\theta$

$$= \int_0^{\pi/2} \int_0^3 \sqrt{9-r^2} \, dr \, d\theta$$

$$= \int_0^{\pi/2} \left[\frac{r}{2}\sqrt{9-r^2} + \frac{9}{2}\sin^{-1}\frac{r}{3}\right]_0^3 d\theta$$

$$= \frac{9\pi}{4} \int_0^{\pi/2} d\theta = \frac{9}{8}\pi^2$$

27. $\int_0^1 \int_0^{\sqrt{1-x^2}} \int_0^2 \sin(x^2+y^2) \, dz \, dy \, dx = \int_0^{\pi/2} \int_0^1 \int_0^2 \sin(r^2)r \, dz \, dr \, d\theta = 2\int_0^{\pi/2} \int_0^1 r\sin(r^2) \, dr \, d\theta$

$$= 2\int_0^{\pi/2} \left[-\frac{1}{2}\cos(r^2)\right]_0^1 d\theta = (1-\cos 1)\int_0^{\pi/2} d\theta = \frac{\pi}{2}(1-\cos 1) \cong 0.7221$$

29. $V = \int_{-\pi/2}^{\pi/2} \int_0^{2a\cos\theta} \int_0^r r \, dz \, dr \, d\theta = \int_{-\pi/2}^{\pi/2} \int_0^{2a\cos\theta} r^2 \, dr \, d\theta$

$$= \int_{-\pi/2}^{\pi/2} \frac{8}{3}a^3\cos^3\theta \, d\theta = \frac{32}{9}a^3$$

31. $V = \int_{-\pi/2}^{\pi/2} \int_0^{a\cos\theta} \int_0^{a-r} r \, dz \, dr \, d\theta = \int_{-\pi/2}^{\pi/2} \int_0^{a\cos\theta} r(a-r) \, dr \, d\theta$

$$= \int_{-\pi/2}^{\pi/2} a^3\left(\frac{1}{2}\cos^2\theta - \frac{1}{3}\cos^3\theta\right) d\theta = \frac{1}{36}a^2(9\pi - 16)$$

33. $V = \int_{-\pi/2}^{\pi/2} \int_0^{\cos\theta} \int_{r^2}^{r\cos\theta} r \, dz \, dr \, d\theta = \int_{-\pi/2}^{\pi/2} \int_0^{\cos\theta} \left(r^2\cos\theta - r^3\right) dr \, d\theta$

$$= \int_{-\pi/2}^{\pi/2} \frac{1}{12}\cos^4\theta \, d\theta = \frac{1}{32}\pi$$

35. $V = \int_0^{2\pi} \int_0^{1/2} \int_{r\sqrt{3}}^{\sqrt{1-r^2}} r \, dz \, dr \, d\theta = \int_0^{2\pi} \int_0^{1/2} \left(r\sqrt{1-r^2} - r^2\sqrt{3}\right) dr \, d\theta = \frac{1}{3}\pi\left(2 - \sqrt{3}\right)$

SECTION 16.10

1. $\left(\sqrt{3},\ \frac{1}{4}\pi,\ \cos^{-1}\left[\frac{1}{3}\sqrt{3}\right]\right)$ 3. $\left(\frac{3}{4},\ \frac{3}{4}\sqrt{3},\ \frac{3}{2}\sqrt{3}\right)$

5. $\rho = \sqrt{2^2 + 2^2 + (2\sqrt{6}/3)^2} = \dfrac{4\sqrt{6}}{3}$ 7. $x = \rho\sin\phi\cos\theta = 3\sin 0\,\cos(\pi/2) = 0$

$\phi = \cos^{-1}\left(\dfrac{2\sqrt{6}/3}{4\sqrt{6}/3}\right) = \cos^{-1}(1/2) = \dfrac{\pi}{3}$ $y = \rho\sin\phi\sin\theta = 3\sin 0\,\sin(\pi/2) = 0$

$\theta = \tan^{-1}(1) = \dfrac{\pi}{4}$ $z = \rho\cos\phi = 3\cos 0 = 3$

$(\rho,\theta,\phi) = \left(\dfrac{4\sqrt{6}}{3},\ \dfrac{\pi}{4},\ \dfrac{\pi}{3}\right)$ $(x,y,z) = (0,0,3)$

9.   The circular cylinder $x^2 + y^2 = 1$;   the radius of the cylinder is 1 and the axis is the z-axis.

11.   The lower nappe of the circular cone $z^2 = x^2 + y^2$.

13.   Horizontal plane one unit above the xy-plane.

15. $\displaystyle\int_0^{\pi/3}\int_0^{2\pi}\int_0^1 \rho^2\sin\phi\,d\rho\,d\theta\,d\phi = \frac{1}{3}\int_0^{\pi/3}\int_0^{2\pi}\sin\phi\,d\theta\,d\phi = \frac{2\pi}{3}\int_0^{\pi/3}\sin\phi\,d\phi = \frac{\pi}{3}$

17. $\displaystyle\int_0^{\pi/4}\int_0^{\pi}\int_0^{2\cos\phi} \rho^2\sin\phi\,d\rho\,d\theta\,d\phi = \frac{8}{3}\int_0^{\pi/4}\int_0^{\pi}\cos^3\phi\,\sin\phi\,d\theta\,d\phi$

$\displaystyle\hspace{6cm} = \frac{8}{3}\pi\int_0^{\pi/4}\cos^3\phi\,\sin\phi\,d\phi$

$\displaystyle\hspace{6cm} = \frac{8}{3}\pi\left[-\frac{1}{4}\cos^4\phi\right]_0^{\pi/4} = \frac{\pi}{2}$

19. $\displaystyle\int_0^1\int_0^{\sqrt{1-x^2}}\int_{\sqrt{x^2+y^2}}^{\sqrt{2-x^2-y^2}} dz\,dy\,dx = \int_{\pi/4}^{\pi/2}\int_0^{\pi/2}\int_0^{\sqrt{2}} \rho^2\sin\phi\,d\rho\,d\theta\,d\phi$

$\displaystyle\hspace{6cm} = \frac{2}{3}\sqrt{2}\int_{\pi/4}^{\pi/2}\int_0^{\pi/2}\sin\phi\,d\theta\,d\phi$

$\displaystyle\hspace{6cm} = \frac{\sqrt{2}}{3}\pi\int_{\pi/4}^{\pi/2}\sin\phi\,d\phi = \frac{\pi}{3}$

21. $\displaystyle\int_0^3\int_0^{\sqrt{9-y^2}}\int_0^{\sqrt{9-x^2-y^2}} z\sqrt{x^2+y^2+x^2}\,dz\,dx\,dy$

$\displaystyle\hspace{1cm} = \int_0^{\pi/2}\int_0^{\pi/2}\int_0^3 \rho\cos\phi\cdot\rho\cdot\rho^2\sin\phi\,d\rho\,d\theta\,d\phi$

$\displaystyle\hspace{1cm} = \int_0^{\pi/2}\frac{1}{2}\sin 2\phi\,d\phi\int_0^{\pi/2}d\theta\int_0^3 \rho^4\,d\rho = \left[-\frac{1}{4}\cos 2\phi\right]_0^{\pi/2}\left(\frac{\pi}{2}\right)\left[\frac{1}{5}\rho^5\right]_0^3$

$\displaystyle\hspace{1cm} = \frac{243\pi}{20}$

23. $V = \int_0^{2\pi} \int_0^{\pi} \int_0^{R} \rho^2 \sin\phi \, d\rho \, d\phi \, d\theta = \frac{4}{3}\pi R^3$

25. $V = \int_0^{\alpha} \int_0^{\pi} \int_0^{R} \rho^2 \sin\phi \, d\rho \, d\phi \, d\theta = \frac{2}{3}\alpha R^3$

27.

$$M = \int_0^{2\pi} \int_0^{\tan^{-1}(r/h)} \int_0^{h\sec\phi} k\rho^3 \sin\phi \, d\rho \, d\phi \, d\theta$$

$$= \int_0^{2\pi} \int_0^{\tan^{-1}(r/h)} \frac{kh^4}{4} \tan\phi \sec^3\phi \, d\phi \, d\theta$$

$$= \frac{kh^4}{4} \int_0^{2\pi} \frac{1}{3}\left[\sec^3\phi\right]_0^{\tan^{-1}(r/h)} d\theta = \frac{kh^4}{4} \int_0^{2\pi} \frac{1}{3}\left[\left(\frac{\sqrt{r^2+h^2}}{h}\right)^3 - 1\right] d\theta$$

$$= \frac{1}{6} k\pi h \left[(r^2 + h^2)^{3/2} - h^3\right]$$

29. center ball at origin;   density $= \dfrac{M}{V} = \dfrac{3M}{4\pi R^3}$

(a) $I = \dfrac{3M}{4\pi R^3} \displaystyle\int_0^{2\pi} \int_0^{\pi} \int_0^{R} \rho^4 \sin^3\phi \, d\rho \, d\phi \, d\theta = \dfrac{2}{5}MR^2$

(b) $I = \frac{2}{5}MR^2 + R^2 M = \frac{7}{5}MR^2$ (parallel axis theorem)

31. center balls at origin;   density $= \dfrac{M}{V} = \dfrac{3M}{4\pi\left(R_2{}^3 - R_1{}^3\right)}$

(a) $I = \dfrac{3M}{4\pi\left(R_2{}^3 - R_1{}^3\right)} \displaystyle\int_0^{2\pi} \int_0^{\pi} \int_{R_1}^{R_2} \rho^4 \sin^3\phi \, d\rho \, d\phi \, d\theta = \dfrac{2}{5}M\left(\dfrac{R_2{}^5 - R_1{}^5}{R_2{}^3 - R_1{}^3}\right)$

This result can be derived from Exercise 29 without further integration. View the solid as a ball of mass M_2 from which is cut out a core of mass M_1.

$$M_2 = \frac{M}{V}V_2 = \frac{3M}{4\pi\left(R_2{}^3 - R_1{}^3\right)}\left(\frac{4}{3}\pi R_2{}^3\right) = \frac{MR_2{}^3}{R_2{}^3 - R_1{}^3}; \quad \text{similarly} \quad M_1 = \frac{MR_1{}^3}{R_2{}^3 - R_1{}^3}.$$

Then

$$I = I_2 - I_1 = \frac{2}{5}M_2 R_2{}^2 - \frac{2}{5}M_1 R_1{}^2 = \frac{2}{5}\left(\frac{MR_2{}^3}{R_2{}^3 - R_1{}^3}\right)R_2{}^2 - \frac{2}{5}\left(\frac{MR_1{}^3}{R_2{}^3 - R_1{}^3}\right)R_1{}^2$$

$$= \frac{2}{5}M\left(\frac{R_2{}^5 - R_1{}^5}{R_2{}^3 - R_1{}^3}\right).$$

(b) Outer radius R and inner radius R_1 gives

$$\text{moment of inertia} = \frac{2}{5}M\left(\frac{R^5 - R_1{}^5}{R^3 - R_1{}^3}\right). \qquad \text{[part}(a)\text{]}$$

As $R_1 \to R$,

$$\frac{R^5 - R_1{}^5}{R^3 - R_1{}^3} = \frac{R^4 + R^3 R_1 + R^2 R_1{}^2 + R R_1{}^3 + R_1{}^4}{R^2 + R R_1 + R_1{}^2} \longrightarrow \frac{5R^4}{3R^2} = \frac{5}{3} R^2.$$

Thus the moment of inertia of spherical shell of radius R is

$$\tfrac{2}{5} M \left(\tfrac{5}{3} R^2 \right) = \tfrac{2}{3} M R^2.$$

(c) $I = \tfrac{2}{3} MR^2 + R^2 M = \tfrac{5}{3} MR^2$ (parallel axis theorem)

33. $V = \displaystyle\int_0^{2\pi} \int_0^{\alpha} \int_0^{a} \rho^2 \sin\phi \, d\rho \, d\phi \, d\theta = \frac{2}{3} \pi \left(1 - \cos\alpha \right) a^3$

35. (a) Substituting $x = \rho \sin\phi \cos\theta, \quad y = \rho \sin\phi \sin\theta, \quad z = \rho \cos\phi$

into $x^2 + y^2 + (z - R)^2 = R^2$

we have $\rho^2 \sin^2\phi + (\rho\cos\phi - R)^2 = R^2,$

which simplifies to $\rho = 2R \cos\phi.$

(b) $0 \le \theta \le 2\pi, \quad 0 \le \phi \le \pi/4, \quad R \sec\phi \le \rho \le 2R\cos\phi$

37. $V = \displaystyle\int_0^{2\pi} \int_0^{\pi/4} \int_0^{2} \rho^2 \sin\phi \, d\rho \, d\phi \, d\theta + \int_0^{2\pi} \int_{\pi/4}^{\pi/2} \int_0^{2\sqrt{2}\,\cos\phi} \rho^2 \sin\phi \, d\rho \, d\phi \, d\theta$

$\qquad = \dfrac{1}{3} \left(16 - 6\sqrt{2} \right) \pi$

39. Encase T in a spherical wedge W. W has spherical coordinates in a box Π that contains S. Define f to be zero outside of T. Then

$$F(\rho, \theta, \phi) = f \left(\rho \sin\phi \cos\theta, \; \rho \sin\phi \sin\theta, \; \rho \cos\phi \right)$$

is zero outside of S and

$$\iiint\limits_{T} f(x, y, z) \, dx\,dy\,dz = \iiint\limits_{W} f(x, y, z) \, dx\,dy\,dz$$

$$= \iiint\limits_{\Pi} F(\rho, \theta, \phi) \, \rho^2 \sin\phi \, d\rho\,d\theta\,d\phi$$

$$= \iiint\limits_{S} F(\rho, \theta, \phi) \, \rho^2 \sin\phi \, d\rho\,d\theta\,d\phi.$$

41. T is the set of all (x, y, z) with spherical coordinates (ρ, θ, ϕ) in the set

$$S: \quad 0 \le \theta \le 2\pi, \quad 0 \le \phi \le \pi/4, \quad R \sec\phi \le \rho \le 2R\cos\phi.$$

T has volume $V = \frac{2}{3} \pi R^3$. By symmetry the $\mathbf{i}, \mathbf{j}$ components of force are zero and

$$\mathbf{F} = \left\{ \frac{3GmM}{2\pi R^3} \iiint\limits_T \frac{z}{(x^2 + y^2 + z^2)^{3/2}} \, dx\,dy\,dz \right\} \mathbf{k}$$

$$= \left\{ \frac{3GmM}{2\pi R^3} \iiint\limits_S \left(\frac{\rho \cos \phi}{\rho^3} \right) \rho^2 \sin \phi \, d\rho\,d\theta\,d\phi \right\} \mathbf{k}$$

$$= \left\{ \frac{3GmM}{2\pi R^3} \int_0^{2\pi} \int_0^{\pi/4} \int_{R \sec \phi}^{2R \cos \phi} \cos \phi \, \sin \phi \, d\rho \, d\phi \, d\theta \right\} \mathbf{k}$$

$$= \frac{GmM}{R^2} \left(\sqrt{2} - 1 \right) \mathbf{k}.$$

SECTION 16.11

1. $ad - bc$ **3.** $2\left(v^2 - u^2\right)$ **5.** $u^2 v^2 - 4uv$

7. abc **9.** r **11.** $w\left(1 + w \cos v\right)$

13. Set $u = x + y$, $v = x - y$. Then

$$x = \frac{u + v}{2}, \quad y = \frac{u - v}{2} \quad \text{and} \quad J(u, v) = -\frac{1}{2}.$$

Ω is the set of all (x, y) with uv-coordinates in

$$\Gamma: \quad 0 \le u \le 1, \quad 0 \le v \le 2.$$

Then

$$\iint\limits_\Omega \left(x^2 - y^2\right) dx\,dy = \iint\limits_\Gamma \frac{1}{2} uv \, du\,dv = \frac{1}{2} \int_0^1 \int_0^2 uv \, dv \, du$$

$$= \frac{1}{2} \left(\int_0^1 u \, du \right) \left(\int_0^2 v \, dv \right) = \frac{1}{2} \left(\frac{1}{2} \right) (2) = \frac{1}{2}.$$

15. $\dfrac{1}{2} \displaystyle\int_0^1 \int_0^2 u \cos(\pi v) \, dv \, du = \dfrac{1}{2} \left(\displaystyle\int_0^1 u \, du \right) \left(\displaystyle\int_0^2 \cos(\pi v) \, dv \right) = \dfrac{1}{2} \left(\dfrac{1}{2} \right) (0) = 0$

17. Set $u = x - y$, $v = x + 2y$. Then

$$x = \frac{2u + v}{3}, \quad y = \frac{v - u}{3}, \quad \text{and} \quad J(u, v) = \frac{1}{3}.$$

Ω is the set of all (x, y) with uv-coordinates in the set

$$\Gamma: 0 \le u \le \pi, \quad 0 \le v \le \pi/2.$$

Therefore

$$\iint_{\Omega} \sin(x-y)\cos(x+2y)\,dxdy = \iint_{\Gamma} \frac{1}{3}\sin u \cos v\,dudv = \frac{1}{3}\int_0^{\pi}\int_0^{\pi/2}\sin u\,\cos v\,dv\,du$$

$$= \frac{1}{3}\left(\int_0^{\pi}\sin u\,du\right)\left(\int_0^{\pi/2}\cos v\,dv\right) = \frac{1}{3}(2)(1) = \frac{2}{3}.$$

19. Set $u = xy, \quad v = y.$ Then

$$x = u/v, \quad y = v \quad\text{and}\quad J(u,v) = 1/v.$$

$xy = 1, \quad xy = 4 \implies u = 1, \quad u = 4$

$y = x, \quad y = 4x \implies u/v = v, \quad 4u/v = v \implies v^2 = u, \quad v^2 = 4u$

Ω is the set of all (x,y) with uv-coordinates in the set

$$\Gamma: \quad 1 \le u \le 4, \quad \sqrt{u} \le v \le 2\sqrt{u}.$$

(a) $A = \iint_{\Gamma} \frac{1}{v}\,dudv = \int_1^4\int_{\sqrt{u}}^{2\sqrt{u}}\frac{1}{v}\,dv\,du = \int_1^4 \ln 2\,du = 3\ln 2$

(b) $\bar{x}A = \int_1^4\int_{\sqrt{u}}^{2\sqrt{u}}\frac{u}{v^2}\,dv\,du = \frac{7}{3}; \quad \bar{x} = \frac{7}{9\ln 2}$

$\bar{y}A = \int_1^4\int_{\sqrt{u}}^{2\sqrt{u}}dv\,du = \frac{14}{3}; \quad \bar{y} = \frac{14}{9\ln 2}$

21. Set $u = x+y, \quad v = 3x - 2y.$ Then

$$x = \frac{2u+v}{5}, \quad y = \frac{3u-v}{5} \quad\text{and}\quad J(u,v) = -\frac{1}{5}.$$

With $\Gamma: \quad 0 \le u \le 1, \quad 0 \le v \le 2$

$$M = \int_0^1\int_0^2 \frac{1}{5}\lambda\,dv\,du = \frac{2}{5}\lambda \quad\text{where}\quad \lambda \text{ is the density.}$$

Then

$$I_x = \int_0^1\int_0^2\left(\frac{3u-v}{5}\right)^2\frac{1}{5}\lambda\,dv\,du = \frac{8\lambda}{375} = \frac{4}{75}\left(\frac{2}{5}\lambda\right) = \frac{4}{75}M,$$

$$I_y = \int_0^1\int_0^2\left(\frac{2u+v}{5}\right)^2\frac{1}{5}\lambda\,dv\,du = \frac{28\lambda}{375} = \frac{14}{75}\left(\frac{2}{5}\lambda\right) = \frac{14}{75}M,$$

$$I_z = I_x + I_y = \tfrac{18}{75}M.$$

23. Set $u = x - 2y, \quad v = 2x + y.$ Then

$$x = \frac{u+2v}{5}, \quad y = \frac{v-2u}{5} \quad\text{and}\quad J(u,v) = \frac{1}{5}.$$

Γ is the region between the parabola $v = u^2 - 1$ and the line $v = 2u + 2$. A sketch of the curves shows that

$$\Gamma: \quad -1 \le u \le 3, \quad u^2 - 1 \le v \le 2u + 2.$$

Then

$$A = \frac{1}{5}\,(\text{area of }\Gamma) = \frac{1}{5}\int_{-1}^{3}\left[(2u+2) - (u^2-1)\right]du = \frac{32}{15}.$$

25. The choice $\theta = \pi/6$ reduces the equation to $13u^2 + 5v^2 = 1$. This is an ellipse in the uv-plane with area $\pi ab = \pi/\sqrt{65}$. Since $J(u,v) = 1$, the area of Ω is also $\pi/\sqrt{65}$.

27. $J = ab\alpha r \cos^{\alpha-1}\theta \sin^{\alpha-1}\theta$

29. $J = abc\rho^2 \sin\phi; \quad V = \int_0^{2\pi}\int_0^{\pi}\int_0^1 abc\rho^2 \sin\phi\, d\rho\, d\phi\, d\theta = \frac{4}{3}\pi abc$

31.
$$V = \frac{2}{3}\pi abc, \quad \lambda = \frac{M}{V} = \frac{3M}{2\pi abc}$$

$$I_x = \frac{3M}{2\pi abc}\int_0^{2\pi}\int_0^{\pi/2}\int_0^1 \left(b^2\rho^2 \sin^2\phi \sin^2\theta + c^2\rho^2 \cos^2\phi\right)abc\rho^2 \sin\phi\, d\rho\, d\phi\, d\theta$$

$$= \tfrac{1}{5}M\left(b^2 + c^2\right)$$

$$I_y = \tfrac{1}{5}M\left(a^2 + c^2\right), \quad I_z = \tfrac{1}{5}M\left(a^2 + b^2\right)$$

33.
$$\iint_{S_a} \frac{e^{-(x-y)^2}}{1+(x+y)^2}\, dx dy = \frac{1}{2}\iint_{\Gamma} \frac{e^{-u^2}}{1+v^2}\, du dv$$

where Γ is the square in the uv-plane with vertices $(-2a, 0)$, $(0, -2a)$, $(2a, 0)$, $(0, 2a)$. Γ contains the square $-a \le u \le a$, $-a \le v \le a$ and is contained in the square $-2a \le u \le 2a$, $-2a \le v \le 2a$.   Therefore

$$\frac{1}{2}\int_{-a}^{a}\int_{-a}^{a} \frac{e^{-u^2}}{1+v^2}\, du\, dv \le \frac{1}{2}\iint_{\Gamma} \frac{e^{-u^2}}{1+v^2}\, du dv \le \frac{1}{2}\int_{-2a}^{2a}\int_{-2a}^{2a} \frac{e^{-u^2}}{1+v^2}\, du\, dv.$$

The two extremes can be written

$$\frac{1}{2}\left(\int_{-a}^{a} e^{-u^2}du\right)\left(\int_{-a}^{a}\frac{1}{1+v^2}dv\right) \quad \text{and} \quad \frac{1}{2}\left(\int_{-2a}^{2a} e^{-u^2}du\right)\left(\int_{-2a}^{2a}\frac{1}{1+v^2}dv\right).$$

As $a \to \infty$ both expressions tend to $\frac{1}{2}\left(\sqrt{\pi}\right)(\pi) = \frac{1}{2}\pi^{3/2}$. It follows that

$$\int_{-\infty}^{\infty}\int_{-\infty}^{\infty} \frac{e^{-(x-y)^2}}{1+(x+y)^2}\, dx\, dy = \frac{1}{2}\pi^{3/2}.$$

PROJECTS AND EXPLORATIONS

16.1.   (a) $\int_1^5 \int_{\sin x}^{e^x} \cos(e^x + y) \, dy \, dx \cong 0.37940379$

(b) Some calculations:

subintervals	trapezoidal estimate	error
2×2	-79.55214932	-79.93
4×4	-30.13375338	-30.513
8×8	-7.21033006	-7.589
16×16	0.48401994	0.1046
32×32	1.01654212	0.6371

Most software has a difficult time with an integral such as this. Another way to proceed is to perform the first integration, reducing the double integral to single integral:

$$\int_1^5 \int_{\sin x}^{e^x} \cos(e^x + y) \, dy \, dx = \int_1^5 \left[\sin(2e^x) - \sin(e^x + \sin x)\right] dx$$

Try the trapezoidal rule (from Chapter 8) on this integral

Now try your trapezoidal rule program for double integrals on

$$\int_1^3 \int_{\sin x}^{e^x} \cos(e^x + y) \, dy \, dx.$$

You should find that your program runs better on this integral.

(c) Handtiming the calculations given in part (b) (program written on a TI-85), we found:

subintervals	time
4×4	5 sec
8×8	11 sec
16×16	33 sec

If most of the time is taken up by the number of points evaluated, we would expect that doubling the number of points would double the time. Here the time seems to be tripling so other parts of the program are taking significant time.

(d) Let $\{a = x_0, x_1, x_2, \ldots, x_k = b\}$ be a regular partition of the interval $[a, b]$. Then

$$J_k = \frac{b-a}{2k}\left[f(x_0) + 2f(x_1) + 2f(x_2) + \cdots + 2f(x_{k-1}) + f(x_k)\right]$$

Now let $\{a = y_0, y_1, y_2, \ldots, y_{2k} = b\}$ be a regular partition of $[a, b]$, then

$$J_{2k} = \frac{b-a}{4k}\left[f(y_0) + 2f(y_1) + 2f(y_2) + \cdots + 2f(y_{2k-1}) + f(y_{2k})\right]$$

where

$$y_0 = x_0, \quad y_1 = \frac{x_0 + x_1}{2}, \quad y_2 = x_1, \quad y_3 = \frac{x_1 + x_2}{2}, \quad y_4 = x_2, \quad \text{and so on}$$

It is easy to verify that

$$\frac{4J_{2k} - J_k}{3} = \left[f(y_0) + 4f(y_1) + 2f(y_2) + 4f(y_3) + 2f(y_2) + \cdots + 4f(y_{2k-1}) + f(y_{2k})\right] = S_k,$$

which is Simpson's rule for the partition $\{a = x_0, x_1, x_2, \ldots, x_k = b\}$. Thus

$$\frac{4J_{2k} - J_k}{3} \cong J.$$

CHAPTER 17

SECTION 17.1

1. (a) $\mathbf{h}(x,y) = y\,\mathbf{i} + x\,\mathbf{j};$ $\mathbf{r}(u) = u\,\mathbf{i} + u^2\,\mathbf{j},$ $u \in [0,1]$

$x(u) = u,$ $y(u) = u^2;$ $x'(u) = 1,$ $y'(u) = 2u$

$\mathbf{h}(\mathbf{r}(u)) \cdot \mathbf{r}'(u) = y(u)\,x'(u) + x(u)\,y'(u) = u^2(1) + u(2u) = 3u^2$

$\displaystyle\int_C \mathbf{h}(\mathbf{r}) \cdot d\mathbf{r} = \int_0^1 3u^2\,du = 1$

(b) $h(x,y) = y\,\mathbf{i} + x\,\mathbf{j};$ $\mathbf{r}(u) = u^3\,\mathbf{i} - 2u\,\mathbf{j},$ $u \in [0,1]$

$x(u) = u^3,$ $y(u) = -2u;$ $x'(u) = 3u^2,$ $y'(u) = -2$

$\mathbf{h}(\mathbf{r}(u)) \cdot \mathbf{r}'(u) = y(u)\,x'(u) + x(u)\,y'(u) = (-2u)(3u^2) + u^3(-2) = -8u^3$

$\displaystyle\int_C \mathbf{h}(\mathbf{r}) \cdot d\mathbf{r} = \int_0^1 -8u^3\,du = -2$

3. $h(x,y) = y\,\mathbf{i} + x\,\mathbf{j};$ $\mathbf{r}(u) = \cos u\,\mathbf{i} - \sin u\,\mathbf{j},$ $u \in [0, 2\pi]$

$x(u) = \cos u,$ $y(u) = -\sin u;$ $x'(u) = -\sin u,$ $y'(u) = -\cos u$

$\mathbf{h}(\mathbf{r}(u)) \cdot \mathbf{r}'(u) = y(u)\,x'(u) + x(u)\,y'(u) = \sin^2 u - \cos^2 u$

$\displaystyle\int_C \mathbf{h}(\mathbf{r}) \cdot d\mathbf{r} = \int_0^{2\pi} (\sin^2 u - \cos^2 u)\,du = 0$

5. (a) $\mathbf{r}(u) = (2 - u)\,\mathbf{i} + (3 - u)\,\mathbf{j},$ $u \in [0,1]$

$\displaystyle\int_C \mathbf{h}(\mathbf{r}) \cdot d\mathbf{r} = \int_0^1 (-5 + 5u - u^2)\,du = -\frac{17}{6}$

(b) $\mathbf{r}(u) = (1 + u)\,\mathbf{i} + (2 + u)\,\mathbf{j},$ $u \in [0,1]$

$\displaystyle\int_C \mathbf{h}(\mathbf{r}) \cdot d\mathbf{r} = \int_0^1 (1 + 3u + u^2)\,du = \frac{17}{6}$

7. $C = C_1 \cup C_2 \cup C_3$ where,

$C_1 : \mathbf{r}(u) = (1 - u)(-2\,\mathbf{i}) + u(2\,\mathbf{i}) = (4u - 2)\,\mathbf{i},$ $u \in [0,1]$

$C_2 : \mathbf{r}(u) = (1 - u)(2\,\mathbf{i}) + u(2\,wj) = (2 - 2u)\,\mathbf{i} + 2u\,\mathbf{j},$ $u \in [0,1]$

$C_3 : \mathbf{r}(u) = (1 - u)(2\,\mathbf{j}) + u(-2\,\mathbf{i}) = -2u\,\mathbf{i} + (2 - 2u)\,\mathbf{j},$ $u \in [0,1]$

$\displaystyle\int_C = \int_{C_1} + \int_{C_2} + \int_{C_3} = 0 + (-4) + (-4) = -8$

9.

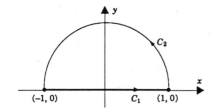

$C_1: \mathbf{r}(u) = (-1 + 2u)\,\mathbf{i}, \quad u \in [0, 1]$

$C_2: \mathbf{r}(u) = \cos u\,\mathbf{i} + \sin u\,\mathbf{j}, \quad u \in [0, \pi]$

$$\int_C = \int_{C_1} + \int_{C_2} = 0 + (-\pi) = -\pi$$

11. (a) $\mathbf{r}(u) = u\,\mathbf{i} + u\,\mathbf{j} + u\,\mathbf{k}, \quad u \in [0, 1]$

$$\int_C \mathbf{h}(\mathbf{r}) \cdot d\mathbf{r} = \int_0^1 3u^2 \, du = 1$$

(b) $\displaystyle\int_C \mathbf{h}(\mathbf{r}) \cdot d\mathbf{r} = \int_0^1 (2u^3 + u^5 + 3u^6) \, du = \frac{23}{21}$

13. (a) $\mathbf{r}(u) = 2u\,\mathbf{i} + 3u\,\mathbf{j} - u\,\mathbf{k}, \quad u \in [0, 1]$

$$\int_C \mathbf{h}(\mathbf{r}) \cdot d\mathbf{r} = \int_0^1 (2\cos 2u + 3\sin 3u + 3u^2) \, du = \left[\sin 2u - \cos 3u + u^3\right]_0^1 = 2 + \sin 2 - \cos 3$$

(b) $\displaystyle\int_C \mathbf{h}(\mathbf{r}) \cdot d\mathbf{r} = \int_0^1 \left(2u\cos u^2 + 3u^2\sin u^3 - u^4\right) du = \left[\sin u^2 - \cos u^3 - \frac{1}{5}u^5\right]_0^1 = \frac{4}{5} + \sin 1 - \cos 1$

15. $\mathbf{r}(u) = (1-u)(\mathbf{j} + 4\mathbf{k}) + u(\mathbf{i} - 4\mathbf{k})$

$= u\,\mathbf{i} + (1-u)\,\mathbf{j} + (4 - 8u)\,\mathbf{k}, \quad u \in [0, 1]$

$$\int_C \mathbf{F}(\mathbf{r}) \cdot d\mathbf{r} = \int_0^1 (-32u + 97u^2 - 64u^3) \, du = \frac{1}{3}$$

17. (a) $\mathbf{r}(u) = u\,\mathbf{i} + u^2\,\mathbf{j}, \quad u \in [0, 2]$

$$\int_C \mathbf{F}(\mathbf{r}) \cdot d\mathbf{r} = \int_0^2 \left[(u+2)u^2 + (2u + u^2)2u\right] du = \int_0^2 \left(3u^3 + 6u^2\right) du = 28$$

(b) $\displaystyle\int_C \mathbf{F}(\mathbf{r}) \cdot d\mathbf{r} = \int_0^{\pi/2} \left[-(\cos u + 2)\sin^2 u + \cos u(2\cos u + \sin u)\right] du$

$$= \int_0^{\pi/2} \left[-\sin^2 u \cos u + \cos 2u + \sin u \cos u\right] du = \frac{1}{6}$$

19. $\mathbf{r}(u) = \cos u\,\mathbf{i} + \sin u\,\mathbf{j} + u\,\mathbf{k}, \quad u \in [0, 2\pi]$

$$\int_C \mathbf{F}(\mathbf{r}) \cdot d\mathbf{r} = \int_0^{2\pi} \left[-\cos^2 u \sin u + \cos^2 u \sin u + u^2\right] du = \int_0^{2\pi} u^2 \, du = \frac{8\pi^3}{3}$$

21.
$$\int_C \mathbf{q} \cdot d\mathbf{r} = \int_a^b [\mathbf{q} \cdot \mathbf{r}'(u)] \, du = \int_a^b \frac{d}{du} [\mathbf{q} \cdot \mathbf{r}(u)] \, du$$

$$= [\mathbf{q} \cdot \mathbf{r}(b)] - [\mathbf{q} \cdot \mathbf{r}(a)]$$

$$= \mathbf{q} \cdot [\mathbf{r}(b) - \mathbf{r}(a)]$$

$$\int_C \mathbf{r} \cdot d\mathbf{r} = \int_a^b [\mathbf{r}(u) \cdot \mathbf{r}'(u)] \, du$$

$$= \frac{1}{2} \int_a^b \|\mathbf{r}\| \, d\|\mathbf{r}\| \quad \text{(see Exercise 57, Section 13.1)}$$

$$= \frac{1}{2} \left(\|\mathbf{r}(b)\|^2 - \|\mathbf{r}(a)\|^2 \right)$$

23. $\displaystyle\int_C \mathbf{f}(\mathbf{r}) \cdot d\mathbf{r} = \int_a^b [\mathbf{f}(\mathbf{r}(u)) \cdot \mathbf{r}'(u)] \, du = \int_a^b [f(u)\mathbf{i} \cdot \mathbf{i}] \, du = \int_a^b f(u) \, du$

25. $E : \mathbf{r}(u) = a \cos u \, \mathbf{i} + b \sin u \, \mathbf{j}, \quad u \in [0, 2\pi]$

$$W = \int_0^{2\pi} \left[\left(-\frac{1}{2} b \sin u\right)(-a \sin u) + \left(\frac{1}{2} a \cos u\right)(b \cos u) \right] du = \int_0^{2\pi} ab \, du = \pi ab$$

If the ellipse is traversed in the opposite direction, then $W = -\pi ab$. In both cases $|W| = \pi ab = $ area of the ellipse.

27. $\mathbf{r}(t) = \alpha t \mathbf{i} + \beta t^2 \mathbf{j} + \gamma t^3 \mathbf{k}$

$\mathbf{r}'(t) = \alpha \mathbf{i} + 2\beta t \mathbf{j} + 3\gamma t^2 \mathbf{k}$

force at time $t = m\mathbf{r}''(t) = m(2\beta \mathbf{j} + 6\gamma t \mathbf{k})$

$$W = \int_0^1 [m(2\beta \mathbf{j} + 6\gamma t \mathbf{k}) \cdot (\alpha \mathbf{i} + 2\beta t \mathbf{j} + 3\gamma t^2 \mathbf{k})] \, dt$$

$$= m \int_0^1 (4\beta^2 t + 18\gamma^2 t^3) \, dt = \left(2\beta^2 + \frac{9}{2}\gamma^2 \right) m$$

29. Take $C : \mathbf{r}(t) = r \cos t \, \mathbf{i} + r \sin t \, \mathbf{j}, \quad t \in [0, 2\pi]$

$$\int_C \mathbf{v}(\mathbf{r}) \cdot d\mathbf{r} = \int_0^{2\pi} [\mathbf{v}(\mathbf{r}(t)) \cdot \mathbf{r}'(t)] \, dt$$

$$= \int_0^{2\pi} [f(x(t), y(t)) \mathbf{r}(t) \cdot \mathbf{r}'(t)] \, dt$$

$$= \int_0^{2\pi} f(x(t), y(t)) [\mathbf{r}(t) \cdot \mathbf{r}'(t)] \, dt = 0$$

since for the circle $\mathbf{r}(t) \cdot \mathbf{r}'(t) = 0$ identically. The circulation is zero.

31. (a) $\mathbf{r}(u) = (1-u)(\mathbf{i}+2\mathbf{k}) + u(\mathbf{i}+3\mathbf{j}+2\mathbf{k}) = \mathbf{i}+3u\mathbf{j}+2\mathbf{k}, \quad u \in [0,1]$.

$$\int_C \mathbf{F}(\mathbf{r}) \cdot d\mathbf{r} = \int_0^1 \frac{9u\mathbf{k}}{(5+9u^2)^{3/2}}\,du = \left[\frac{-\mathbf{k}}{\sqrt{5+9u^2}}\right]_0^1 = \frac{\mathbf{k}}{\sqrt{5}} - \frac{\mathbf{k}}{\sqrt{14}}$$

(b) $C = C_1 \cup C_2$, where

$C_1 : \mathbf{r}(u) = (1-u)\mathbf{i} + 5u\mathbf{i} = (1+4u)\mathbf{i}, \quad u \in [0,1]$,

$C_2 : x^2 + y^2 + z^2 = 25 \implies \|\mathbf{r}\| = 5$

$$\int_{C_1} \mathbf{F}(\mathbf{r}) \cdot d\mathbf{r} = \int_0^1 \frac{4\mathbf{k}(1+4u)}{(1+4u)^3}\,du = \int_0^1 \frac{4\mathbf{k}}{(1+4u)^2}\,du \left[\frac{-\mathbf{k}}{1+4u}\right]_0^1 = \frac{4}{5}\mathbf{k}$$

$$\int_{C_2} \mathbf{F}(\mathbf{r}) \cdot d\mathbf{r} = \int_{C_2} \frac{k\mathbf{r}}{\|\mathbf{r}\|^3} \cdot d\mathbf{r}$$
$$= \frac{k}{5^3}\int_{C_2} \mathbf{r} \cdot d\mathbf{r} = \frac{k}{5^3}\int_{C_2} \|\mathbf{r}\| d\|\mathbf{r}\| \quad \text{(see Exercise 57, Section 13.1)}$$
$$= \frac{k}{5^3}\left[\frac{1}{2}\|\mathbf{r}\|^2\right]_{(5,0,0)}^{(0,5/\sqrt{2},5/\sqrt{2})} = 0$$

Therefore, $\displaystyle\int_C \mathbf{F}(\mathbf{r}) \cdot d\mathbf{r} = \frac{4}{5}\mathbf{k}$.

33. $\mathbf{r}(u) = u\mathbf{i} + \alpha u(1-u)\mathbf{j}, \quad \mathbf{r}'(u) = \mathbf{i} + \alpha(1-2u)\mathbf{j}, \quad u \in [0,1]$

$$W(\alpha) = \int_C \mathbf{F}(\mathbf{r}) \cdot d\mathbf{r} = \int_0^1 \left[(\alpha^2 u^2(1-u)^2 + 1) + [u + \alpha u(1-u)]\alpha(1-2u)\right]dx$$
$$= \int_0^1 \left[1 + (\alpha+\alpha^2)u - (2\alpha+2\alpha^2)u^2 + \alpha^2 u^4\right]du = 1 - \frac{1}{6}\alpha + \frac{1}{30}\alpha^2$$

$$W'(\alpha) = -\frac{1}{6} + \frac{1}{15}\alpha \implies \alpha = \frac{15}{6}$$

The work done by $\mathbf{F}$ is a minimum when $\alpha = 15/6$.

SECTION 17.2

1. $\mathbf{h}(x,y) = \nabla f(x,y)$ where $f(x,y) = \frac{1}{2}(x^2+y^2)$

C is closed $\implies \displaystyle\int_C \mathbf{h}(\mathbf{r}) \cdot d\mathbf{r} = 0$

3. $\mathbf{h}(x,y) = \nabla f(x,y)$ where $f(x,y) = x\cos\pi y; \quad \mathbf{r}(0) = \mathbf{0}, \quad \mathbf{r}(1) = \mathbf{i}-\mathbf{j}$

$$\int_C \mathbf{h}(\mathbf{r}) \cdot d\mathbf{r} = \int_C \nabla f(\mathbf{r}) \cdot d\mathbf{r} = f(\mathbf{r}(1)) - f(\mathbf{r}(0)) = f(1,-1) - f(0,0) = -1$$

5. $\mathbf{h}(x,y) = \nabla f(x,y)$ where $f(x,y) = \frac{1}{2}x^2 y^2; \quad \mathbf{r}(0) = \mathbf{j}, \quad \mathbf{r}(1) = -\mathbf{j}$

$$\int_C \mathbf{h}(\mathbf{r}) \cdot d\mathbf{r} = \int_C \nabla f(\mathbf{r}) \cdot d\mathbf{r} = f(\mathbf{r}(1)) - f(\mathbf{r}(0)) = f(0,-1) - f(0,1) = 0 - 0 = 0$$

7. $\mathbf{h}(x,y) = \nabla f(x,y)$ where $f(x,y) = x^2 y - xy^2;$ $\mathbf{r}(0) = \mathbf{i},\ \mathbf{r}(\pi) = -\mathbf{i}$

$$\int_C \mathbf{h}(\mathbf{r}) \cdot d\mathbf{r} = \int_C \nabla f(\mathbf{r}) \cdot d\mathbf{r} = f(\mathbf{r}(\pi)) - f(\mathbf{r}(0)) = f(-1,0) - f(1,0) = 0 - 0 = 0$$

9. $\mathbf{h}(x,y) = \nabla f(x,y)$ where $f(x,y) = (x^2 + y^4)^{3/2}$

$$\int_C \mathbf{h}(\mathbf{r}) \cdot d\mathbf{r} = \int_C \nabla f(\mathbf{r}) \cdot d\mathbf{r} = f(1,0) - f(-1,0) = 1 - 1 = 0$$

11. $\mathbf{h}(x,y)$ is not a gradient, but part of it,

$$2x \cosh y\, \mathbf{i} + (x^2 \sinh y - y)\mathbf{j},$$

is a gradient. Since we are integrating over a closed curve, the contribution of the gradient part is 0. Thus

$$\int_C \mathbf{h}(\mathbf{r}) \cdot d\mathbf{r} = \int_C (-y\mathbf{i}) \cdot d\mathbf{r}.$$

$C_1: \mathbf{r}(u) = \mathbf{i} + (-1 + 2u)\mathbf{j}, \quad u \in [0,1]$

$C_2: \mathbf{r}(u) = (1 - 2u)\mathbf{i} + \mathbf{j}, \quad u \in [0,1]$

$C_3: \mathbf{r}(u) = -\mathbf{i} + (1 - 2u)\mathbf{j}, \quad u \in [0,1]$

$C_4: \mathbf{r}(u) = (-1 + 2u)\mathbf{i} - \mathbf{j}, \quad u \in [0,1]$

$$\int_C \mathbf{h}(\mathbf{r}) \cdot d\mathbf{r} = \int_{C_1} (-y\mathbf{i}) \cdot d\mathbf{r} + \int_{C_2} (-y\mathbf{i}) \cdot d\mathbf{r} + \int_{C_3} (-y\mathbf{i}) \cdot d\mathbf{r} + \int_{C_4} (-y\mathbf{i}) \cdot d\mathbf{r}$$

$$= \quad 0 \quad + \int_0^1 -\mathbf{i} \cdot (-2\mathbf{i})\, du + \quad 0 \quad + \int_0^1 \mathbf{i} \cdot (2\mathbf{i})\, du$$

$$= \quad 0 \quad + \quad \int_0^1 2\, du \quad + \quad 0 \quad + \quad \int_0^1 2\, du$$

$$= \quad 4$$

13. $\mathbf{h}(x,y) = (3x^2 y^3 + 2x)\,\mathbf{i} + (3x^3 y^2 - 4y)\,\mathbf{j};$ $\dfrac{\partial P}{\partial y} = 9x^2 y^2 = \dfrac{\partial Q}{\partial x}.$ Thus $\mathbf{h}$ is a gradient.

(a) $\mathbf{r}(u) = u\mathbf{i} + e^u\,\mathbf{j}, \quad \mathbf{r}'(u) = \mathbf{i} + e^u\,\mathbf{j}, \quad u \in [0,1]$

$$\int_C \mathbf{h}(\mathbf{r}) \cdot d\mathbf{r} = \int_0^1 \left[(3u^2 e^{3u} + 2u) + 3u^3 e^3 u - 4e^{2u} \right] du = \left[u^3 e^{3u} + u^2 - 2e^{2u} \right]_0^1 = e^3 - 2e^2 + 3$$

(b) $\dfrac{\partial f}{\partial x} = 3x^2 y^3 + 2x \quad \Longrightarrow \quad f(x,y) = x^3 y^3 + x^2 + g(y);$

$$\dfrac{\partial f}{\partial y} = 3x^3 y^2 + g'(y) = 3x^3 - 4y \quad \Longrightarrow \quad g'(y) = -4y \quad \Longrightarrow \quad g(y) = -2y^2$$

Therefore, $f(x,y) = x^3 y^3 + x^2 - 2y^2.$

Now, at $u = 0$, $r(0) = 0\mathbf{i} + \mathbf{j} = (0,1)$; at $u = 1$, $r(1) = \mathbf{i} + e\mathbf{j} = (1, e)$ and

$$\int_C \mathbf{h}(\mathbf{r}) \cdot d\mathbf{r} = \left[x^3 y^3 + x^2 - 2y^2\right]_{(0,1)}^{(1,e)} = e^3 - 2e^2 + 3$$

15. $\mathbf{h}(x,y) = (e^{2y} - 2xy)\mathbf{i} + (2xe^{2y} - x^2 + 1)\mathbf{j}$; $\dfrac{\partial P}{\partial y} = 2e^{2y} - 2x = \dfrac{\partial Q}{\partial x}$. Thus $\mathbf{h}$ is a gradient.

(a) $\mathbf{r}(u) = ue^u\mathbf{i} + (1+u)\mathbf{j}$, $\mathbf{r}'(u) = (1 = u)e^u\mathbf{i} + \mathbf{j}$, $u \in [0,1]$

$$\int_C \mathbf{h}(\mathbf{r}) \cdot d\mathbf{r} = \int_0^1 \left[e^2(3ue^{3u} + e^{3u} - 2u^3 e^{2u} - 5u^2 e^{2u} - 2ue^{2u} + 1\right] du$$

$$= \left[e^2 ue^{3u} - u^3 e^{2u} - u^2 e^{2u} + u\right]_0^1 = e^5 - 2e^2 + 1$$

(b) $\dfrac{\partial f}{\partial x} = e^{2y} - 2xy \implies f(x,y) = xe^{2y} - x^2 y + g(y)$.

$\dfrac{\partial f}{\partial y} = 2xe^{2y} - x^2 + g'(y) = 3x^3 - 4y \implies g'(y) = 1 \implies g(y) = y$

Therefore, $f(x,y) = xe^{2y} - x^2 y + y$.

Now, at $u = 0$, $r(0) = 0\mathbf{i} + \mathbf{j} = (0,1)$; at $u = 1$, $r(1) = e\mathbf{i} + 2\mathbf{j} = (e, 2)$ and

$$\int_C \mathbf{h}(\mathbf{r}) \cdot d\mathbf{r} = \left[xe^{2y} - x^2 y + y\right]_{(0,1)}^{(e,2)} = e^5 - 2e^2 + 1$$

17. $\mathbf{h}(x,y,z) = (2xz + \sin y)\mathbf{i} + x\cos y\,\mathbf{j} + x^2\mathbf{k}$;

$\dfrac{\partial P}{\partial y} = \cos y = \dfrac{\partial Q}{\partial x}$, $\dfrac{\partial P}{\partial z} = 2x = \dfrac{\partial R}{\partial x}$, $\dfrac{\partial Q}{\partial z} = 0 = \dfrac{\partial R}{\partial y}$. Thus $\mathbf{h}$ is a gradient.

$$\dfrac{\partial f}{\partial x} = 2xz + \sin y, \longrightarrow f(x,y,z) = x^2 z + x\sin y + g(y,z)$$

$\dfrac{\partial f}{\partial y} = x\cos y + \dfrac{\partial g}{\partial y} = x\cos y, \implies g(y,z) = h(z) \implies f(x,y,z) = x^2 z + x\sin y + h(z)$

$$\dfrac{\partial f}{\partial z} = x^2 + h'(z) = x^2 \implies h'(z) = 0 \implies h(z) = C$$

Therefore, $f(x,y,z) = x^2 z + x\sin y$ (take $C = 0$)

$$\int_C \mathbf{h}(\mathbf{r}) \cdot d\mathbf{r} = \int_C \nabla f \cdot d\mathbf{r} = \left[x^2 z + x\sin y\right]_{\mathbf{r}(0)}^{\mathbf{r}(2\pi)} = \left[x^2 z + x\sin y\right]_{(1,0,0)}^{(1,0,2\pi)} = 2\pi$$

19. $\mathbf{F}(x,y) = (x + e^{2y})\mathbf{i} + (2y + 2xe^{2y})\mathbf{j}$; $\dfrac{\partial P}{\partial y} = 2e^{2y} = \dfrac{\partial Q}{\partial x}$. Thus $\mathbf{F}$ is a gradient.

$$\dfrac{\partial f}{\partial x} = x + e^{2y} \implies f(x,y) = \dfrac{1}{2}x^2 + xe^{2y} + g(y);$$

$\dfrac{\partial f}{\partial y} = 2xe^{2y} + g'(y) = 2y + 2xe^{2y} \implies g'(y) = 2y \implies g(y) = y^2$ take $C = 0$

Therefore, $f(x,y) = \dfrac{1}{2}x^2 + xe^{2y} + y^2$ (take $C = 0$).

$$\int_C \mathbf{F}(\mathbf{r}) \cdot d\mathbf{r} = \int_C \nabla f \cdot d\mathbf{r} = \left[\dfrac{1}{2}x^2 + xe^{2y} + y^2\right]_{\mathbf{r}(0)}^{\mathbf{r}(2\pi)} = \left[\dfrac{1}{2}x^2 + xe^{2y} + y^2\right]_{(3,0)}^{(3,0)} = 0$$

21. Set $f(x, y, z) = g(x)$ and $C : \mathbf{r}(u) = u\mathbf{i}, \quad u \in [a, b]$.

In this case

$$\nabla f(\mathbf{r}(u)) = g'(x(u))\mathbf{i} = g'(u)\mathbf{i} \quad \text{and} \quad \mathbf{r}'(u) = \mathbf{i},$$

so that

$$\int_C \nabla f(\mathbf{r}) \cdot d\mathbf{r} = \int_a^b [\nabla f(\mathbf{r}(u)) \cdot \mathbf{r}'(u)] \, du = \int_a^b g'(u) \, du.$$

Since $f(\mathbf{r}(b)) - f(\mathbf{r}(a)) = g(b) - g(a),$

$$\int_C \nabla f(\mathbf{r}) \cdot d\mathbf{r} = f(\mathbf{r}(b)) - f(\mathbf{r}(a)) \quad \text{gives} \quad \int_a^b g'(u) \, du = g(b) - g(a).$$

23. $\mathbf{F}(\mathbf{r}) = \nabla \left(\dfrac{mG}{r} \right); \quad W = \displaystyle\int_C \mathbf{F}(\mathbf{r}) \cdot d\mathbf{r} = mG \left(\dfrac{1}{r_2} - \dfrac{1}{r_1} \right)$

25. $\mathbf{F}(x, y, z) = 0\mathbf{i} + 0\mathbf{j} + \dfrac{-mGr_0^2}{(r_0 + z)^2}\mathbf{k}; \quad \dfrac{\partial P}{\partial y} = 0 = \dfrac{\partial Q}{\partial x}, \quad \dfrac{\partial P}{\partial z} = 0 = \dfrac{\partial R}{\partial x}, \quad \dfrac{\partial Q}{\partial z} = 0 = \dfrac{\partial R}{\partial y}.$

Therefore, $\mathbf{F}(x, y, z)$ is a gradient.

$\dfrac{\partial f}{\partial x} = 0 \implies f(x, y, z) = g(y, z); \quad \dfrac{\partial f}{\partial y} = \dfrac{\partial g}{\partial y} = 0 \implies g(y, z) = h(z).$

Therefore $f(x, y, z) = h(z).$

Now $\dfrac{\partial f}{\partial z} = h'(z) = \dfrac{-mGr_0^2}{(r_0 + z)^2} \implies f(x, y, z) = h(z) = \dfrac{mGr_0^2}{r_0 + z}$

27. By Exercise 25, the work required to lift an object of mass m a distance of h miles above the surface of the earth is:

$$W = \int_0^h -\mathbf{F} \cdot d\mathbf{r} = -f(0, 0, h) + f(0, 0, 0) = \dfrac{mGr_0^2 h}{r_0(r_0 + h)}$$

where $\mathbf{F} = \dfrac{-mGr_0^2}{(r_0 + z)^2}\mathbf{k}, \quad \mathbf{r} = 0\mathbf{i} + 0\mathbf{j} + u\mathbf{k}, \quad u \in [0, h], \quad \text{and} \quad f = \dfrac{mGr_0^2}{r_0 + z}.$

In this particular case, put $r_0 = 4000, \quad h = 500/5280$ and $m = 8000/32 = 250.$

Then $W \cong 23.67\, G.$

SECTION * 17.3

1. If f is continuous, then $-f$ is continuous and has antiderivatives u. The scalar fields $U(x, y, z) = u(x)$ are potential functions for $\mathbf{F}$:

$$\nabla U = \dfrac{\partial U}{\partial x}\mathbf{i} + \dfrac{\partial U}{\partial y}\mathbf{j} + \dfrac{\partial U}{\partial z}\mathbf{k} = \dfrac{du}{dx}\mathbf{i} = -f\mathbf{i} = -\mathbf{F}.$$

3. The scalar field $U(x, y, z) = cz + d$ is a potential energy function for **F**. We know that the total mechanical energy remains constant. Thus, for any times t_1 and t_2,

$$\tfrac{1}{2}m[v(t_1)]^2 + U(\mathbf{r}(t_1)) = \tfrac{1}{2}m[v(t_2)]^2 + U(\mathbf{r}(t_2)).$$

This gives

$$\tfrac{1}{2}m[v(t_1)]^2 + cz(t_1) + d = \tfrac{1}{2}m[v(t_2)]^2 + cz(t_2) + d.$$

Solve this equation for $v(t_2)$ and you have the desired formula.

5. (a) We know that $-\nabla U$ points in the direction of maximum decrease of U. Thus $\mathbf{F} = -\nabla U$ attempts to drive objects toward a region where U has lower values.

    (b) At a point where u has a minimum, $\nabla U = \mathbf{0}$ and therefore $\mathbf{F} = \mathbf{0}$.

7. (a) By conservation of energy $\tfrac{1}{2}mv^2 + U = E$. Since E is constant and U is constant, v is constant.

    (b) ∇U is perpendicular to any surface where U is constant. Obviously so is $\mathbf{F} = -\nabla U$.

9. $f(x, y, z) = -\dfrac{k}{\sqrt{x^2 + y^2 + z^2}}$ is a potential function for **F**. The work done by **F** moving an object along C is:

$$W = \int_C \mathbf{F}(\mathbf{r}) \cdot d\mathbf{r} = \int_a^b \nabla f \cdot d\mathbf{r} = f[\mathbf{r}(b)] - f[\mathbf{r}(a)].$$

Since $\mathbf{r}(a) = (x_0, y_0, z_0)$ and $\mathbf{r}(b) = (x_1, y_1, z_1)$ are points on the unit sphere,

$$f[\mathbf{r}(b)] - f[\mathbf{r}(a)] = -k \quad \text{and so} \quad W = 0$$

SECTION 17.4

1. $\mathbf{r}(u) = u\,\mathbf{i} + 2u\,\mathbf{j}, \quad u \in [0, 1]$

$$\int_C (x - 2y)\,dx + 2x\,dy = \int_0^1 \{[x(u) - 2y(u)]x'(u) + 2x(u)\,y'(u)\}\,du = \int_0^1 u\,du = \frac{1}{2}$$

3. $C = C_1 \cup C_2$

$C_1 : \mathbf{r}(u) = u\,\mathbf{i}, \quad u \in [0, 1]; \qquad C_2 : \mathbf{r}(u) = \mathbf{i} + 2u\,\mathbf{j}, \quad u \in [0, 1]$

$$\int_{C_1} (x - 2y)\,dx + 2x\,dy = \int_{C_1} x\,dx = \int_0^1 x(u)\,x'(u)\,du = \int_0^1 u\,du = \frac{1}{2}$$

$$\int_{C_2} (x - 2y)\,dx + 2x\,dy = \int_{C_2} 2x\,dy = \int_0^1 4\,du = 4$$

$$\int_C = \int_{C_1} + \int_{C_2} = 4\tfrac{1}{2}$$

5. $\mathbf{r}(u) = 2u^2\,\mathbf{i} + u\,\mathbf{j}, \quad u \in [0, 1]$

$$\int_C y\,dx + xy\,dy = \int_0^1 [y(u)\,x'(u) + x(u)\,y(u)\,y'(u)]\,du = \int_0^1 (4u^2 + 2u^3)\,du = \frac{11}{6}$$

7. $C = C_1 \cup C_2$

$$C_1 : \mathbf{r}(u) = u\mathbf{j}, \quad u \in [0,1]; \qquad C_2 : \mathbf{r}(u) = 2u\mathbf{i} + \mathbf{j}, \quad u \in [0,1]$$

$$\int_{C_1} y\,dx + xy\,dy = 0$$

$$\int_{C_2} y\,dx + xy\,dy = \int_{C_2} y\,dx = \int_0^1 y(u)\,x'(u)\,du = \int_0^1 2\,du = 2$$

$$\int_C = \int_{C_1} + \int_{C_2} = 2$$

9. $\mathbf{r}(u) = 2u\,\mathbf{i} + 4u\,\mathbf{j}, \quad u \in [0,1]$

$$\int_C y^2\,dx + (xy - x^2)\,dy = \int_0^1 \left\{ y^2(u)x'(u) + [x(u)y(u) - x^2(u)]\,y'(u) \right\}\,du$$

$$= \int_0^1 \left[(4u)^2(2) + (8u^2 - 4u^2)(4) \right]\,du = \int_0^1 48u^2\,du = 16$$

11. $\mathbf{r}(u) = \frac{1}{8}u^2\mathbf{i} + u\mathbf{j}, \quad u \in [0,4]$

$$\int_C y^2\,dx + (xy - x^2)\,dy = \int_0^4 \left\{ y^2(u)x'(u) + [x(u)y(u) - x^2(u)]\,y'(u) \right\}\,du$$

$$= \int_0^4 \left[u^2\left(\frac{u}{4}\right) + \left(\frac{u^2}{8}(u) - \left(\frac{u^2}{8}\right)^2 (1)\right) \right]\,du$$

$$= \int_0^1 \left[\frac{3}{8}u^3 - \frac{1}{64}u^4 \right]\,du = \frac{104}{5}$$

13. $\mathbf{r}(u) = (1-u)\mathbf{i} + u\mathbf{j}, \quad u \in [0,1]$

$$\int_C x^2y\,dx + xy\,dy = \int_0^1 [x^2(u)y(u)x'(u) + x(u)y(u)y'(u)]\,du$$

$$= \int_0^1 [(1-u)^2 u(-1) + (1-u)u]\,du = \int_0^1 (u^2 - u^3)\,du = \frac{1}{12}$$

15. $\mathbf{r}(u) = \cos u\,\mathbf{i} + \sin u\,\mathbf{j}, \quad u \in [0,\pi/2]$

$$\int_C x^2y\,dx + xy\,dy = \int_0^1 [x^2(u)y(u)x'(u) + x(u)y(u)y'(u)]\,du$$

$$= \int_0^1 [\cos^2 u \sin u(-\sin u) + \cos u \sin u \cos u]\,du$$

$$= \int_0^1 \sin^4 u\,du - \int_0^1 \sin^2 u\,du + \int_0^1 \cos^2 u \sin u\,du$$

$$= \frac{1}{3} - \frac{\pi}{16}$$

17. $\mathbf{r}(u) = u\mathbf{i} + u\mathbf{j}, \quad u \in [0,1]$

$$\int_C (y^2 + 2x + 1)\,dx + (2xy + 4y - 1)\,dy$$

$$= \int_0^1 \left\{ [y^2(u) + 2x(u) + 1]x'(u) + [2x(u)y(u) + 4y(u) - 1]y'(u) \right\}\,du$$

$$\int_0^1 \left[(u^2 + 2u + 1) + (2u^2 + 4u - 1) \right]\,du = \int_0^1 (3u^2 + 6u)\,du = 4$$

19. $\mathbf{r}(u) = u\mathbf{i} + u^3\mathbf{j}, \quad u \in [0,1]$

$$\int_C (y^2 + 2x + 1)\,dx + (2xy + 4y - 1)\,dy$$

$$= \int_0^1 \left\{ [y^2(u) + 2x(u) + 1]x'(u) + [2x(u)y(u) + 4y(u) - 1]y'(u) \right\}\,du$$

$$= \int_0^1 \left[(u^6 + 2u + 1) + (2u^4 + 4u^3 - 1)3u^2 \right]\,du = \int_0^1 \left(7u^6 + 12u^5 - 3u^2 + 2u + 1 \right)\,du = 4$$

21. $\mathbf{r}(u) = u\mathbf{i} + u\mathbf{j} + u\mathbf{k}, \quad u \in [0,1]$

$$\int_C y\,dx + 2z\,dy + x\,dz = \int_0^1 [y(u)\,x'(u) + 2z(u)\,y'(u) + x(u)\,z'(u)]\,du = \int_0^1 4u\,du = 2$$

23. $C = C_1 \cup C_2 \cup C_3$

$C_1: \mathbf{r}(u) = u\mathbf{k}, \quad u \in [0,1]; \quad C_2: \mathbf{r}(u) = u\mathbf{j} + \mathbf{k}, \quad u \in [0,1]; \quad C_3: \mathbf{r}(u) = u\mathbf{i} + \mathbf{j} + \mathbf{k}, \quad u \in [0,1]$

$$\int_{C_1} y\,dx + 2z\,dy + x\,dz = 0$$

$$\int_{C_2} y\,dx + 2z\,dy + x\,dz = \int_{C_2} 2z\,dy = \int_0^1 2z(u)\,y'(u)\,du = \int_0^1 2\,du = 2$$

$$\int_{C_3} y\,dx + 2z\,dy + x\,dz = \int_{C_3} y\,dx = \int_0^1 y(u)\,x'(u)\,du = \int_0^1 du = 1$$

$$\int_C = \int_{C_1} + \int_{C_2} + \int_{C_3} = 3$$

25. $\mathbf{r}(u) = 2u\,\mathbf{i} + 2u\,\mathbf{j} + 8u\,\mathbf{k}, \quad u \in [0,1]$

$$\int_C xy\,dx + 2z\,dy + (y+z)\,dz$$

$$= \int_0^1 \{x(u)y(u)x'(u) + 2z(u)y'(u) + [y(u) + z(u)]z'(u)\}\,du$$

$$= \int_0^1 [(2u)(2u)(2) + 2(8u)(2) + (2u + 8u)(8)]\,du$$

$$= \int_0^1 (8u^2 + 112u)\,du = \frac{176}{3}$$

27. $\mathbf{r}(u) = u\mathbf{i} + u\mathbf{j} + 2u^2\mathbf{k}, \quad u \in [0,2]$

$$\int_C xy\,dx + 2z\,dy + (y+z)\,dz$$

$$= \int_0^2 \{x(u)y(u)x'(u) + 2z(u)y'(u) + [y(u) + z(u)]z'(u)\}\,du$$

$$= \int_0^2 [(u)(u)(1) + 2(2u^2)(1) + (u + 2u^2)(4u)]\,du$$

$$= \int_0^2 (8u^3 + 9u^2)\,du = 56$$

29. $\mathbf{r}(u) = (u-1)\mathbf{i} + (1 + 2u^2)\mathbf{j} + u\mathbf{k}, \quad u \in [1,2]$

$$\int_C x^2 y\,dx + y\,dy + xz\,dz$$

$$= \int_1^2 [x^2(u)y(u)x'(u) + y(u)y'(u) + x(u)z(u)z'(u)]\,du$$

$$= \int_1^2 [(u-1)^2(1 + 2u^2)(1) + (1 + 2u^2)(4u) + (u - 1)u]\,du$$

$$= \int_1^2 (2u^4 + 4u^3 + 4u^2 + u + 1)\,du = \frac{1177}{30}$$

31. (a) $\dfrac{\partial P}{\partial y} = 6x - 4y = \dfrac{\partial Q}{\partial x}$

$$\frac{\partial f}{\partial x} = x^2 + 6xy - 2y^2 \quad \Longrightarrow \quad f(x,y) = \frac{1}{3}x^3 + 3x^2 y - 2xy^2 + g(y)$$

$$\frac{\partial f}{\partial y} = 3x^2 - 4xy + g'(y) = 3x^2 - 4xy + 2y \quad \Longrightarrow \quad g'(y) = 2y \quad \Longrightarrow \quad g(y) = y^2 + C$$

Therefore, $f(x,y) = \dfrac{1}{3}x^3 + 3x^2 y - 2xy^2 + y^2$ (take $C = 0$)

(b) $\displaystyle\int_C (x^2 + 6xy - 2y^2)\,dx + (3x^2 - 4xy + 2y)\,dy = [f(x,y)]_{(3,0)}^{(0,4)} = 7$

(c) $\int_C' (x^2 + 6xy - 2y^2)\,dx + (3x^2 - 4xy + 2y)\,dy = [f(x,y)]_{(4,0)}^{(0,3)} = -\dfrac{37}{3}$

33. $s'(u) = \sqrt{[x'(u)]^2 + [y'(u)]^2} = a$

(a) $M = \displaystyle\int_C k(x+y)\,ds = k \int_0^{\pi/2} [x(u) + y(u)]\,s'(u)\,du = ka^2 \int_0^{\pi/2} (\cos u + \sin u)\,du = 2ka^2$

(b) $\displaystyle x_M M = \int_C kx(x+y)\,ds = k \int_0^{\pi/2} x(u)\,[x(u) + y(u)]\,s'(u)\,du$

$$= ka^3 \int_0^{\pi/2} (\cos^2 u + \cos u \sin u)\,du = \frac{1}{4}ka^3(\pi + 2)$$

$$y_M M = \int_C ky(x+y)\,ds = k \int_0^{\pi/2} y(u)\,[x(u) + y(u)]\,s'(u)\,du$$

$$= ka^3 \int_0^{\pi/2} (\sin u \cos u + \sin^2 u)\,du = \frac{1}{4}ka^3(\pi + 2)$$

$x_M = y_M = \frac{1}{8}a(\pi + 2)$

35. (a) $I_z = \displaystyle\int_C k(x+y)a^2\,ds = a^2 \int_C k(x+y)\,ds = a^2 M = Ma^2$

(b) The distance from the point (x,y) to the line $y = x$ is $|y - x|/\sqrt{2}$.   Therefore

$$I = \int_C k(x+y)\left[\frac{1}{2}(y-x)^2\right]\,ds = \frac{1}{2}k \int_0^{\pi/2} (a\cos u + a\sin u)(a\sin u - a\cos u)^2 a\,du$$

$$= \frac{1}{2}ka^4 \int_0^{\pi/2} (\sin u - \cos u)^2 \frac{d}{du}(\sin u - \cos u)\,du$$

$$= \frac{1}{2}ka^4 \left[\frac{1}{3}(\sin u - \cos u)^3\right]_0^{\pi/2} = \frac{1}{3}ka^4.$$

From Exercise 33, $M = 2ka^2$.   Therefore

$$I = \tfrac{1}{6}(2ka^2)a^2 = \tfrac{1}{6}Ma^2.$$

37. (a) $s'(u) = \sqrt{a^2 + b^2}$

$$L = \int_C ds = \int_0^{2\pi} \sqrt{a^2 + b^2}\,du = 2\pi\sqrt{a^2 + b^2}$$

(b) $x_M = 0, \quad y_M = 0 \quad$ (by symmetry)

$$z_M = \frac{1}{L}\int_C z\,ds = \frac{1}{2\pi\sqrt{a^2+b^2}} \int_0^{2\pi} bu\sqrt{a^2 + b^2}\,du = b\pi$$

(c) $I_x = \displaystyle\int_C \frac{M}{L}(y^2 + z^2)\,ds = \frac{M}{2\pi} \int_0^{2\pi} (a^2 \sin^2 u + b^2 u^2)\,du = \frac{1}{6}M(3a^2\pi + 8b^2\pi^2)$

$I_y = \frac{1}{6}M(3a^2\pi + 8b^2\pi^2)$ similarly

$I_z = Ma^2$ (all the mass is at distance a from the z-axis)

39.
$$M = \int_C k(x^2 + y^2 + z^2)\,ds$$
$$= k\sqrt{a^2 + b^2}\int_0^{2\pi}(a^2 + b^2u^2)\,du = \frac{2}{3}\pi k\sqrt{a^2+b^2}\,(3a^2 + 4\pi^2b^2)$$

SECTION 17.5

1. (a) $\oint_C xy\,dx + x^2\,dy = \int_{C_1} xy\,dx + x^2\,dy + \int_{C_2} xy\,dx + x^2\,dy + \int_{C_3} xy\,dx + x^2\,dy,$ where

$C_1: \mathbf{r}(u) = u\mathbf{i} + u\mathbf{j},\ u\in[0,1];\quad C_2: \mathbf{r}(u) = (1-u\mathbf{i}+\mathbf{j},\ u\in[0,1]$

$C_3: \mathbf{r}(u) = (1-u)\mathbf{j},\ u\in[0,1].$

$$\int_{C_1} xy\,dx + x^2\,dy = \int_0^1(u^2 + u^2)\,du = \frac{2}{3}$$

$$\int_{C_2} xy\,dx + x^2\,dy = \int_0^1 -(1-u)\,du = -\frac{1}{2}$$

$$\int_{C_3} xy\,dx + x^2\,dy = \int_0^1 0^2(-1)\,du = 0$$

Therefore, $\oint_C xy\,dx + x^2\,dy = \frac{2}{3} - \frac{1}{2} = \frac{1}{6}.$

(b) $\oint_C xy\,dx + x^2\,dy = \iint_\Omega x\,dx\,dy = \int_0^1\int_0^y x\,dx\,dy = \int_0^1\left[\frac{1}{2}x^2\right]_0^y du = \frac{1}{2}\int_0^1 y^2\,dy = \frac{1}{6}$

3. (a) $C: \mathbf{r}(u) = \cos u\,\mathbf{i} + \sin u\,\mathbf{j},\quad u\in[0,2\pi]$

$$\oint_C(3x^2 + y)\,dx + (2x + y^3)\,dy\int_0^{2\pi}[(3\cos^2 u + \sin u)(-\sin u) + (2\cos u + \sin^3 u)\cos u]\,du$$

$$= \int_0^{2\pi}[3\cos^2 u(-\sin u) - \sin^2 u + 2\cos^2 u + \sin^3 u\,\cos u]\,du$$

$$= \left[\cos^3 u - \frac{1}{2}u + \frac{1}{4}\sin 2u + u + \frac{1}{2}\sin 2u + \frac{1}{4}\sin^4 u\right]_0^{2\pi} = \pi$$

(b) $\oint_C(3x^2 + y)\,dx + (2x + y^3)\,dy = \iint_\Omega 1\,dx\,dy = \text{area }\Omega = \pi$

5. $\oint_C 3y\,dx + 5x\,dy = \iint_\Omega(5-3)\,dx\,dy = 2A = 2\pi$

7. $\oint_C x^2\,dy = \iint_\Omega 2x\,dx\,dy = 2\bar{x}A = 2\left(\frac{a}{2}\right)(ab) = a^2b$

9. $\displaystyle\oint_C (3xy + y^2)\, dx + (2xy + 5x^2)\, dy = \iint_\Omega [(2y + 10x) - (3x + 2y)]\, dx\, dy$

$\displaystyle = \iint_\Omega 7x\, dx\, dy = 7\,\overline{x}A = 7(1)(\pi) = 7\pi$

11. $\displaystyle\oint_C (2x^2 + xy - y^2)\, dx + (3x^2 - xy + 2y^2)\, dy = \iint_\Omega [(6x - y) - (x - 2y)]\, dx\, dy$

$\displaystyle = \iint_\Omega (5x + y)\, dx\, dy = (5\overline{x} + \overline{y})A = (5a + 0)(\pi r^2) = 5a\pi r^2$

13. $\displaystyle\oint_C e^x \sin y\, dx + e^x \cos y\, dy = \iint_\Omega [e^x \cos y - e^x \cos y]\, dx\, dy = 0$

15. $\displaystyle\oint_C 2xy\, dx + x^2\, dy = \iint_\Omega [2x - 2x]\, dx\, dy = 0$

17. $C: \mathbf{r}(u) = a \cos u\, \mathbf{i} + a \sin u\, \mathbf{j}; \quad u \in [0, 2\pi]$

$\displaystyle A = \oint_C -y\, dx = \int_0^{2\pi} (-a \sin u)(-\sin u)\, du = a^2 \int_0^{2\pi} \sin^2 u\, du = a^2 \left[\frac{1}{2}u - \frac{1}{4}\sin 2u\right]_0^{2\pi} = \pi a^2$

19. $A = \displaystyle\oint_C x\, dy,$ where $C = C_1 \cup C_2 \cup C_3;$

$C_1: \mathbf{r}(u) = au\, \mathbf{i}, \quad u \in [0,1]; \quad C_2: \mathbf{r}(u) = a(1-u)\mathbf{i} + bu\, \mathbf{j}, \quad u \in [0,1];$

$C_3: \mathbf{r}(u) = b(1-u)\mathbf{j}, \quad u \in [0,1].$

$\displaystyle\int_{C_1} x\, dy = 0; \quad \int_{C_2} x\, dy = \int_0^1 ab(1-u)\, du = \frac{1}{2}ab; \quad \int_{C_3} x\, dy = 0$

Therefore, $A = \dfrac{1}{2}ab.$

21. $\displaystyle\oint_C (ay + b)\, dx + (cx + d)\, dy = \iint_\Omega (c - a)\, dx\, dy = (c - a)A$

23. We take the arch from $x = 0$ to $x = 2\pi R$. (Figure 9.11.1) Let C_1 be the line segment from $(2\pi R, 0)$ and let C_2 be the cycloidal arch from $(2\pi R, 0)$ back to $(0,0)$. Letting $C = C_1 \cup C_2$,

$$A = \oint_C x\,dy = \int_{C_1} x\,dy + \int_{C_2} x\,dy = 0 + \int_{C_2} x\,dy$$

$$= \int_{2\pi}^{0} R(\theta - \sin\theta)(R\sin\theta)\,d\theta$$

$$= R^2 \int_0^{2\pi} (\sin^2\theta - \theta\sin\theta)\,d\theta$$

$$= R^2\left[\frac{\theta}{2} - \frac{\sin 2\theta}{4} + \theta\cos\theta - \sin\theta\right]_0^{2\pi} = 3\pi R^2.$$

25. $\mathbf{F}(x,y) = (x^2 - y^3)\mathbf{i} + (x^2 + y^2)\mathbf{j};$ $C: \mathbf{r} = \cos u\,\mathbf{i} + \sin u\,\mathbf{j},$ $u \in [0, 2\pi]$

$$W = \oint_C \mathbf{F}\cdot d\mathbf{r} = \iint_\Omega (2x + 3y^2)\,dx\,dy = \int_0^{2\pi}\int_0^1 (2r\cos\theta + 3r^2\sin^2\theta)\,r\,dr\,d\theta$$

$$= \int_0^{2\pi}\int_0^1 (2r^2\cos\theta + 3r^3\sin^2\theta)\,dr\,d\theta$$

$$= \frac{2}{3}\int_0^{2\pi}\cos\theta\,d\theta + \frac{3}{4}\int_0^{2\pi}\sin^2\theta\,d\theta$$

$$= \frac{2}{3}[\sin\theta]_0^{2\pi} + \frac{3}{4}\left[\frac{1}{2}u - \frac{1}{4}\sin 2u\right]_0^{2\pi} = \frac{3}{4}\pi$$

27. Taking Ω to be of type II (see Figure 17.5.2), we have

$$\iint_\Omega \frac{\partial Q}{\partial x}(x,y)\,dx\,dy = \int_c^d \int_{\phi_3(y)}^{\phi_4(y)} \frac{\partial Q}{\partial x}(x,y)\,dx\,dy$$

$$= \int_c^d \{Q[\phi_4(y), y] - Q[\phi_3(y), y]\}\,dy$$

$$(*) = \int_c^d Q[\phi_4(y), y]\,dy - \int_c^d Q[\phi_3(y), y]\,dy.$$

graph of $x = \phi_4(y)$ from $x = c$ to $x = d$ is the curve

$$C_4: \mathbf{r}_4(u) = \phi_4(u)\mathbf{i} + u\mathbf{j}, \qquad u \in [c, d].$$

h of $x = \phi_3(y)$ from $x = c$ to $x = d$ is the curve

$$C_3: \mathbf{r}_3(u) = \phi_3(u)\mathbf{i} + u\mathbf{j}, \qquad u \in [c, d].$$

$$\oint_C Q(x,y)\,dy = \int_{C_4} Q(x,y)\,dy - \int_{C_3} Q(x,y)\,dy$$

$$= \int_c^d Q[\phi_4(u), u]\,du - \int_c^d Q[\phi_3(u), u]\,du.$$

$0,0)$ to we have

le, it can be replaced by y. Comparison with $(*)$ gives the result.

29. $\oint_{C_1} = \oint_{C_2} + \oint_{C_3}$

31. $\dfrac{\partial P}{\partial y} = \dfrac{-2xy}{(x^2+y^2)^2} = \dfrac{\partial Q}{\partial x}$ except at $(0,0)$

(a) If C does not enclose the origin, and Ω is the region enclosed by C, then

$$\oint_C \frac{x}{x^2+y^2}\,dx + \frac{y}{x^2+y^2}\,dy = \iint_\Omega 0\,dxdy = 0.$$

(b) If C does enclose the origin, then

$$\oint_C = \oint_{C_a}$$

where $C_a: \mathbf{r}(u) = a\cos u\,\mathbf{i} + a\sin u\,\mathbf{j}, \quad u \in [0,2\pi]$ is a small circle in the inner region of C.

In this case

$$\oint_C = \int_0^{2\pi}\left[\frac{a\cos u}{a^2}(-a\sin u) + \frac{a\sin u}{a^2}(a\cos u)\right]du = \int_0^{2\pi} 0\,du = 0.$$

The integral is still 0.

33. If Ω is the region enclosed by C, then

$$\oint_C \mathbf{v}\cdot d\mathbf{r} = \oint_C \frac{\partial\phi}{\partial x}\,dx + \frac{\partial\phi}{\partial y}\,dy = \iint_\Omega \left\{\frac{\partial}{\partial x}\left(\frac{\partial\phi}{\partial y}\right) - \frac{\partial}{\partial y}\left(\frac{\partial\phi}{\partial x}\right)\right\}dxdy$$

$$= \iint_\Omega 0\,dxdy = 0.$$

equality of mixed partials

35. $A = \dfrac{1}{2}\oint_C(-y\,dx + x\,dy)$

$$= \left[\int_{C_1} + \int_{C_2} + \cdots \int_{C_n}\right]$$

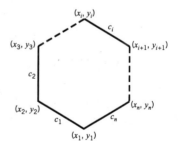

Now

$$\int_{C_i}(-y\,dx + x\,dy) = \int_0^1\{[y_i + u(y_{i+1}-y_i)](x_{i+1}-x_i) + [x_i + u(x_{i+1}-x_i)](y_{i+1}-y_i)\}\,du$$

$$= x_iy_{i+1} - x_{i+1}y_i, \quad i=1,2,\ldots,n; \ x_{n+1}=x_1, \ y_{n+1}=y_1$$

Thus, $A = \dfrac{1}{2}\left[(x_1y_2 - x_2y_1) + (x_2y_3 - x_3y_2) + \cdots + (x_ny_1 - x_1y_n)\right]$

SECTION 17.6

1. $4[(u^2 - v^2)\mathbf{i} - (u^2 + v^2)\mathbf{j} + 2uv\mathbf{k}]$ 3. $2(\mathbf{j} - \mathbf{i})$

5. $\mathbf{r}(u,v) = 3\cos u \cos v\,\mathbf{i} + 2\sin u \cos v\,\mathbf{j} + 6\sin v\,\mathbf{k}, \quad u \in [0\,2\pi], \; v \in [0, \pi/2]$

7. $\mathbf{r}(u,v) = 2\cos u \cos v\,\mathbf{i} + 2\sin u \cos v\,\mathbf{j} + 2\sin v\,\mathbf{k}, \quad u \in [0\,2\pi], \; v \in (\,\pi/4, \pi/2\,]$

9.   The surface consists of all points of the form $(x, g(x,z), z)$ with $(x,z) \in \Omega$. This set of points

     is given by
$$\mathbf{r}(u,v) = u\mathbf{i} + g(u,v)\mathbf{j} + v\mathbf{k}, \quad (u,v) \in \Omega.$$

11. $x^2/a^2 + y^2/b^2 + z^2/c^2 = 1;$ ellipsoid

13. $x^2/a^2 - y^2/b^2 = z;$ hyperbolic paraboloid

15.   For each $v \in [a,b]$, the points on the surface at level $z = f(v)$ form a circle of radius v. That circle can be parametrized

$$\mathbf{R}(u) = v\cos u\,\mathbf{i} + v\sin u\,\mathbf{j} + f(v)\mathbf{k}, \quad u \in [0, 2\pi].$$

Letting v range over $[a,b]$, we obtain the entire surface:

$$\mathbf{r}(u,v) = v\cos u\,\mathbf{i} + v\sin u\,\mathbf{j} + f(v)\mathbf{k}; \quad 0 \le u \le 2\pi, \quad a \le v \le b.$$

17.   Since γ is the angle between p and the xy-plane, γ is the angle between the upper normal to p and $\mathbf{k}$. (Draw a figure.) Therefore, by 17.6.5,

$$\text{area of } \Gamma = \iint_\Omega \sec\gamma \, dxdy = (\sec\gamma)A_\Omega = A_\Omega \sec\gamma.$$

$\qquad\qquad\qquad\qquad\qquad\quad$ └── γ is constant

19.   The surface is the graph of the function

$$f(x,y) = c\left(1 - \frac{x}{a} - \frac{y}{b}\right) = \frac{c}{ab}(ab - bx - ay)$$

defined over the triangle $\Omega : 0 \le x \le a, \quad 0 \le y \le b(1 - x/a)$. Note that Ω has area $\frac{1}{2}ab$.

$$A = \iint_{\Omega} \sqrt{[f'_x(x,y)]^2 + [f'_y(x,y)]^2 + 1} \; dx dy$$

$$= \iint_{\Omega} \sqrt{c^2/a^2 + c^2/b^2 + 1} \; dx dy$$

$$= \frac{1}{ab} \sqrt{a^2 b^2 + a^2 c^2 + b^2 c^2} \iint_{\Omega} dx dy = \frac{1}{2} \sqrt{a^2 b^2 + a^2 c^2 + b^2 c^2}.$$

21. $f(x,y) = x^2 + y^2, \quad \Omega : 0 \leq x^2 + y^2 \leq 4$

$$A = \iint_{\Omega} \sqrt{4x^2 + 4y^2 + 1} \; dx dy \qquad [\,\text{change to polar coordinates}\,]$$

$$= \int_0^{2\pi} \int_0^2 \sqrt{4r^2 + 1} \, r \, dr \, d\theta$$

$$= 2\pi \left[\tfrac{1}{12}(4r^2 + 1)^{3/2} \right]_0^2 = \tfrac{1}{6}\pi(17\sqrt{17} - 1)$$

23. $f(x,y) = a^2 - (x^2 + y^2), \quad \Omega : \tfrac{1}{4}a^2 \leq x^2 + y^2 \leq a^2$

$$A = \iint_{\Omega} \sqrt{4x^2 + 4y^2 + 1} \; dx dy \qquad [\,\text{change to polar coordinates}\,]$$

$$= \int_0^{2\pi} \int_{a/2}^a r\sqrt{4r^2 + 1} \, dr \, d\theta = 2\pi \left[\frac{1}{12}(4r^2 + 1)^{3/2} \right]_{a/2}^a$$

$$= \frac{\pi}{6} \left[(4a^2 + 1)^{3/2} - (a^2 + 1)^{3/2} \right]$$

25. $f(x,y) = \tfrac{1}{3}(x^{3/2} + y^{3/2}), \quad \Omega : 0 \leq x \leq 1, \quad 0 \leq y \leq x$

$$A = \iint_{\Omega} \frac{1}{2}\sqrt{x + y + 4} \; dx dy$$

$$= \int_0^1 \int_0^x \frac{1}{2}\sqrt{x + y + 4} \; dy \, dx = \int_0^1 \left[\frac{1}{3}(x + y + 4)^{3/2} \right]_0^x dx$$

$$= \int_0^1 \frac{1}{3} \left[(2x + 4)^{3/2} - (x + 4)^{3/2} \right] dx = \frac{1}{3} \left[\frac{1}{5}(2x + 4)^{5/2} - \frac{2}{5}(x + 4)^{5/2} \right]_0^1$$

$$= \tfrac{1}{15}(36\sqrt{6} - 50\sqrt{5} + 32)$$

27. The surface $x^2 + y^2 + z^2 - 4z = 0$ is a sphere of radius 2 centered at $(0,0,2)$:

$$x^2 + y^2 + z^2 - 4z = 0 \quad \Longleftrightarrow \quad x^2 + y^2 + (z - 2)^2 = 4.$$

The quadric cone $z^2 = 3(x^2 + y^2)$ intersects the sphere at height $z = 3$:

$$\left.\begin{array}{r}x^2 + y^2 + z^2 - 4z = 0\\ z^2 = 3(x^2 + y^2)\end{array}\right\} \quad \Longrightarrow \quad \begin{array}{r}3(x^2 + y^2) + 3z^2 - 12z = 0\\ 4z^2 - 12z = 0\\ z = 3. \quad (\text{since } z \geq 2)\end{array}$$

The surface of which we are asked to find the area is a spherical segment of width 1 (from $z = 3$ to $z = 4$) in a sphere of radius 2. The area of the segment is 4π. (Exercise 27, Section 9.10.)

A more conventional solution. The spherical segment is the graph of the function

$$f(x,y) = 2 + \sqrt{4 - (x^2 + y^2)}, \quad \Omega : 0 \leq x^2 + y^2 \leq 3.$$

Therefore

$$A = \iint_\Omega \sqrt{\left(\frac{-x}{\sqrt{4 - x^2 - y^2}}\right)^2 + \left(\frac{-y}{\sqrt{4 - x^2 - y^2}}\right)^2 + 1} \; dxdy$$

$$= \iint_\Omega \frac{2}{\sqrt{4 - (x^2 + y^2)}} \; dxdy$$

$$= \int_0^{2\pi} \int_0^{\sqrt{3}} \frac{2r}{\sqrt{4 - r^2}} \; dr \, d\theta \qquad [\text{changed to polar coordinates}]$$

$$= 2\pi \left[-2\sqrt{4 - r^2}\right]_0^{\sqrt{3}} = 4\pi$$

29. (a) $\displaystyle \iint_\Omega \sqrt{\left[\frac{\partial g}{\partial y}(y,z)\right]^2 + \left[\frac{\partial g}{\partial z}(y,z)\right]^2 + 1} \; dydz = \iint_\Omega \sec\left[\alpha(y,z)\right] \, dydz$

where α is the angle between the unit normal with positive $\mathbf{i}$ component and the positive x-axis

(b) $\displaystyle \iint_\Omega \sqrt{\left[\frac{\partial h}{\partial x}(x,z)\right]^2 + \left[\frac{\partial h}{\partial z}(x,z)\right]^2 + 1} \; dxdz = \iint_\Omega \sec\left[\beta(x,z)\right] \, dxdz$

where β is the angle between the unit normal with positive $\mathbf{j}$ component and the positive y-axis

31. (a) $\mathbf{N}(u,v) = v \cos u \sin \alpha \cos \alpha \, \mathbf{i} + v \sin u \sin \alpha \cos \alpha \, \mathbf{j} - v \sin^2 \alpha \, \mathbf{k}$

(b) $$A = \iint_\Omega \|\mathbf{N}(u,v)\| \, dudv = \iint_\Omega v \sin \alpha \, dudv$$

$$= \int_0^{2\pi} \int_0^s v \sin \alpha \, dv \, du = \pi s^2 \sin \alpha$$

33. $A = \sqrt{A_1^2 + A_2^2 + A_3^2}$; the unit normal to the plane of Ω is a vector of the form

$$\cos\gamma_1\,\mathbf{i}+\cos\gamma_2\,\mathbf{j}+\cos\gamma_3\,\mathbf{k}.$$

Note that

$$A_1 = A\cos\gamma_1, \quad A_2 = A\cos\gamma_2, \quad A_3 = A\cos\gamma_3.$$

Therefore

$$A_1{}^2 + A_2{}^2 + A_3{}^2 = A^2[\cos^2\gamma_1 + \cos^2\gamma_2 + \cos^2\gamma_3] = A^2.$$

35. (a) (We use Exercise 34.) $f(r,\theta) = r + \theta; \quad \Omega : 0 \le r \le 1, \quad 0 \le \theta\pi$

$$A = \iint\limits_{\Omega} \sqrt{r^2\,[f_r'(r,\theta)]^2 + [f_\theta'(r,\theta)]^2 + r^2}\,\,drd\theta = \iint\limits_{\Omega} \sqrt{2r^2+1}\,\,drd\theta$$

$$= \int_0^\pi \int_0^1 \sqrt{2r^2+1}\,dr\,d\theta = \tfrac{1}{4}\sqrt{2}\,\pi\left[\sqrt{6} + \ln\left(\sqrt{2}+\sqrt{3}\right)\right]$$

(b) $f(r,\theta) = re^\theta; \quad \Omega : 0 \le r \le a, \quad 0 \le \theta \le 2\pi$

$$A = \iint\limits_{\Omega} r\sqrt{2e^{2\theta}+1}\,\,drd\theta = \left(\int_0^{2\pi}\sqrt{2e^{2\theta}+1}\,d\theta\right)\left(\int_0^a r\,dr\right)$$

$$= \tfrac{1}{2}a^2[\sqrt{2e^{4\pi}+1} - \sqrt{3} + \ln\left(1+\sqrt{3}\right) - \ln\left(1+\sqrt{2e^{4\pi}+1}\right)]$$

SECTION 17.7

For Exercises 1–5 we have $\sec\left[\gamma(x,y)\right] = \sqrt{y^2+1}$.

1. $\displaystyle\iint\limits_{S} d\sigma = \int_0^1\int_0^1 \sqrt{y^2+1}\,dx\,dy = \int_0^1 \sqrt{y^2+1}\,dy = \tfrac{1}{2}[\sqrt{2}+\ln(1+\sqrt{2})]$

3. $\displaystyle\iint\limits_{S} 3y\,d\sigma = \int_0^1\int_0^1 3y\sqrt{y^2+1}\,dy\,dx = \int_0^1 3y\sqrt{y^2+1}\,dy = \left[(y^2+1)^{3/2}\right]_0^1 = 2\sqrt{2}-1$

5. $\displaystyle\iint\limits_{S}\sqrt{2z}\,d\sigma = \iint\limits_{S} y\,d\sigma = \tfrac{1}{3}(2\sqrt{2}-1)$ (Exercise 3)

7. $\displaystyle\iint\limits_{S} xy\,d\sigma; \quad S: \mathbf{r}(u,v) = (6-2u-3v)\,\mathbf{i} + u\,\mathbf{j} + v\,\mathbf{k}, \quad 0 \le u \le 3 - \tfrac{3}{2}v, \quad 0 \le v \le 2$

$$\|\mathbf{N}(u,v)\| = \|(-2\mathbf{i}+\mathbf{j})\times(-3\mathbf{i}+\mathbf{k})\| = \sqrt{14}$$

$$\iint\limits_S xy\, d\sigma = \sqrt{14} \iint\limits_\Omega x(u,v)y(u,v)\, du\, dv$$

$$= \sqrt{14} \iint\limits_\Omega (6 - 2u - 3v)u\, du\, dv$$

$$= \sqrt{14} \int_0^2 \int_0^{3-3v/2} (6u - 2u^2 - 3uv)\, du\, dv$$

$$= \sqrt{14} \left[3\left(3 - \tfrac{3}{2}v\right)^2 - \tfrac{2}{3}\left(3 - \tfrac{3}{2}v\right)^3 - \tfrac{3}{2}v\left(3 - \tfrac{3}{2}v\right)^2 \right] dv = \frac{9}{2}\sqrt{14}$$

9. $\displaystyle\iint\limits_S x^2 z\, d\sigma;\quad S: \mathbf{r}(u,v) = (\cos u\,\mathbf{i} + v\,\mathbf{j} + \sin u\,\mathbf{k}, \quad 0 \le u \le \pi, \quad 0 \le v \le 2.$

$$\mathbf{N}(u,v) = \begin{vmatrix} \mathbf{i} & \mathbf{j} & \mathbf{k} \\ -\sin u & 0 & \cos u \\ 0 & 1 & 0 \end{vmatrix} = -\cos u\,\mathbf{i} - \sin u\,\mathbf{k} \quad \text{and} \quad \|\mathbf{N}(u,v)\| = 1.$$

$$\iint\limits_S x^2 z\, d\sigma = \iint\limits_\Omega \cos^2 u\, \sin u\, du = \int_0^2 \int_0^\pi \cos^2 u\, \sin u\, du = \frac{4}{3}$$

11. $\displaystyle\iint\limits_S (x^2 + y^2)\, d\sigma;\quad S: \mathbf{r}(u,v) = (\cos u \cos v\,\mathbf{i} + \cos u \sin v\,\mathbf{j} + \sin u\,\mathbf{k}, \quad 0 \le u \le \pi/2, \quad 0 \le v \le 2\pi.$

$$\mathbf{N}(u,v) = \begin{vmatrix} \mathbf{i} & \mathbf{j} & \mathbf{k} \\ -\sin u \cos v & -\sin u \sin v & \cos u \\ -\cos u \cos v & \cos u \cos v 1 & 0 \end{vmatrix} = -\cos^2 u \cos v\,\mathbf{i} + \cos^2 u \sin v\, w\mathbf{j} - \sin u \cos u\,\mathbf{k};$$

$$\|\mathbf{N}(u,v)\| = \cos u.$$

$$\iint\limits_S (x^2 + y^2)\, d\sigma = \iint\limits_\Omega \cos^2 u \cos u\, du = \int_0^{2\pi} \int_0^{\pi/2} \cos^3 u\, du = \frac{4}{3}\pi$$

For Exercises 13–15 the surface S is given by

$$f(x,y) = a - x - y; \quad 0 \le x \le a, \quad 0 \le y \le a - x.$$

Then $\quad \sec[\gamma(x,y)] = \sqrt{3}.$

13. $\displaystyle M = \iint\limits_S \lambda(x,y,x)\, d\sigma = \int_0^a \int_0^{a-x} k\sqrt{3}\, dy\, dx = \int_0^a k\sqrt{3}(a-x)\, dx = \frac{1}{2}a^2 k\sqrt{3}$

15. $\displaystyle M = \iint\limits_S \lambda(x,y,z)\, d\sigma = \int_0^a \int_0^{a-x} kx^2\sqrt{3}\, dy\, dx = \int_0^a k\sqrt{3}x^2(a-x)\, dx = \frac{1}{12}a^4 k\sqrt{3}$

17. $\quad S: \mathbf{r}(u,v) = a\cos u \cos v\,\mathbf{i} + a\sin u \cos v\,\mathbf{j} + a\sin v\,\mathbf{k}\quad$ with $\quad 0 \le u \le 2\pi, \quad 0 \le v \le \tfrac{1}{2}\pi.$ By a previous calculation $\quad \|\mathbf{N}(u,v)\| = a^2 \cos v.$

$$\bar{x} = 0, \quad \bar{y} = 0 \quad \text{(by symmetry)}$$

$$\bar{z}A = \iint\limits_{S} z \, d\sigma = \iint\limits_{\Omega} z(u,v)\, \|\mathbf{N}(u,v)\|\, dudv = \int_0^{2\pi}\int_0^{\pi/2} a^3 \sin v \cos v \, dv \, du = \pi a^3$$

$$\bar{z} = \tfrac{1}{2}a \quad \text{since} \quad A = 2\pi a^2$$

19. $\mathbf{N}(u,v) = (\mathbf{i}+\mathbf{j}+2\mathbf{k})\cdot(\mathbf{i}-\mathbf{j}) = 2\mathbf{i}+2\mathbf{j}-2\mathbf{k}$

$$\text{flux in the direction of } \mathbf{N} = \iint\limits_{S}\left(\mathbf{v}\cdot\frac{\mathbf{N}}{\|\mathbf{N}\|}\right)d\sigma = \iint\limits_{\Omega} [\mathbf{v}(x(u),\,y(u),\,z(u))\cdot\mathbf{N}(u,v)]\, dudv$$

$$= \iint\limits_{\Omega} [(u+v)\mathbf{i}-(u-v)\mathbf{j}]\cdot[2\mathbf{i}+2\mathbf{j}-2\mathbf{k}]\, dudv.$$

$$= \iint\limits_{\Omega} 4v \, dudv = 4\int_0^1\int_0^1 v\, dv\, du = 2$$

For Exercises 21–23 $\mathbf{n} = \dfrac{1}{a}(x\mathbf{i}+y\mathbf{j}+z\mathbf{k})$

$S : \mathbf{r}(u,v) = a\cos u\cos v\,\mathbf{i}+a\sin u\cos v\,\mathbf{j}+a\sin v\,\mathbf{k}$ with $0\le u\le 2\pi,\quad -\tfrac{1}{2}\pi \le v\le \tfrac{1}{2}\pi$

$\|\mathbf{N}(u,v)\| = a^2\cos v$

21.   with $\mathbf{v} = z\mathbf{k}$

$$\text{flux} = \iint\limits_{S}(\mathbf{v}\cdot\mathbf{n})\,d\sigma = \frac{1}{a}\iint\limits_{S} z^2\, d\sigma = \frac{1}{a}\iint\limits_{\Omega}(a^2\sin^2 v)(a^2\cos v)\, dudv$$

$$= a^3\int_0^{2\pi}\int_{-\pi/2}^{\pi/2}(\sin^2 v\cos v)\,d\sigma = \frac{4}{3}\pi a^3$$

23.   with $\mathbf{v} = y\mathbf{i}-x\mathbf{j}$

$$\text{flux} = \iint\limits_{S}(\mathbf{v}\cdot\mathbf{n})\,d\sigma = \frac{1}{a}\iint\limits_{S}\underbrace{(yx-xy)}_{0}\,d\sigma = 0$$

For Exercises 25–27 the triangle S is the graph of the function

$$f(x,y) = a-x-y \quad \text{on} \quad \Omega : 0\le x\le a,\quad 0\le y\le a-x.$$

The triangle has area $A = \tfrac{1}{2}\sqrt{3}a^2$.

25.   with $\mathbf{v} = x\mathbf{i}+y\mathbf{j}+z\mathbf{k}$

$$\text{flux} = \iint\limits_{S}(\mathbf{v}\cdot\mathbf{n})\,d\sigma = \iint\limits_{\Omega}(-v_1 f_x' - v_2 f_y' + v_3)\, dxdy$$

$$= \iint\limits_{\Omega}[-x(-1)-y(-1)+(a-x-y)]\, dxdy = a\iint\limits_{\Omega} dxdy = aA = \frac{1}{2}\sqrt{3}a^3$$

27. with $v = x^2\mathbf{i} - y^2\mathbf{j}$

$$\text{flux} = \iint\limits_{S} (\mathbf{v} \cdot \mathbf{n})\, d\sigma = \iint\limits_{\Omega} (-v_1 f_x' - v_2 f_y' + v_3)\, dx\, dy$$

$$= \iint\limits_{\Omega} [-x^2(-1) - (-y^2)(-1) + 0]\, dx\, dy = \int_0^a \int_0^{a-x} (x^2 - y^2)\, dy\, dx$$

$$= \int_0^a \left[ax^2 - x^3 - \frac{1}{3}(a-x)^3 \right] dx = \left[\frac{1}{3}ax^3 - \frac{1}{4}x^4 + \frac{1}{12}(a-x)^4 \right]_0^a = 0$$

29.

$$\text{flux} = \iint\limits_{S} (\mathbf{v} \cdot \mathbf{n})\, d\sigma = \iint\limits_{\Omega} (-v_1 f_x' - v_2 f_y' + v_3)\, dx\, dy$$

$$= \iint\limits_{\Omega} (-x^3 y - xy)\, dx\, dy = \int_0^1 \int_0^2 (-x^3 y - xy)\, dy\, dx$$

$$= \int_0^1 -2(x^3 + x)\, dx = -\frac{3}{2}$$

31. $\mathbf{n} = \dfrac{1}{a}(x\mathbf{i} + y\mathbf{j})$

$$\text{flux} = \iint\limits_{S} (\mathbf{v} \cdot \mathbf{n})\, d\sigma = \frac{1}{a} \iint\limits_{S} [(x\mathbf{i} + y\mathbf{j} + z\mathbf{k}) \cdot (x\mathbf{i} + y\mathbf{j})]\, d\sigma$$

$$= \frac{1}{a} \iint\limits_{S} (x^2 + y^2)\, d\sigma = a \iint\limits_{S} d\sigma = a\,(\text{area of } S) = a\,(2\pi a l) = 2\pi a^2 l$$

33.

$$\text{flux} = \iint\limits_{S} (\mathbf{v} \cdot \mathbf{n})\, d\sigma = \iint\limits_{\Omega} (-v_1 f_x' - v_2 f_y' + v_3)\, dx\, dy = \iint\limits_{\Omega} 2y^{3/2}\, dx\, dy$$

$$= \int_0^1 \int_0^{1-x} 2y^{3/2}\, dy\, dx = \int_0^1 \frac{4}{5}(1-x)^{5/2}\, dx = \frac{8}{35}$$

35.

$$\text{flux} = \iint\limits_{S} (\mathbf{v} \cdot \mathbf{n})\, d\sigma = \iint\limits_{\Omega} (-v_1 f_x' - v_2 f_y' + v_3)\, dx\, dy = \iint\limits_{\Omega} -y^{5/2}\, d\sigma$$

$$= \int_0^1 \int_0^{1-x} -y^{5/2}\, dy\, dx = \int_0^1 -\frac{2}{7}(1-x)^{7/2}\, dx = -\frac{4}{63}$$

37. $\bar{x} = 0, \quad \bar{y} = 0 \qquad$ by symmetry

verify that $\|\mathbf{N}(u, v)\| = v \sin \alpha$

$$\bar{z}A = \iint\limits_{S} z\,d\sigma = \iint\limits_{\Omega} (v\cos\alpha)(v\sin\alpha)\,du\,dv = \sin\alpha\cos\alpha \int_{0}^{2\pi}\!\!\int_{0}^{s} v^2\,dv\,du = \tfrac{2}{3}\pi\sin\alpha\cos\alpha\,s^3$$

$$\bar{z} = \tfrac{2}{3}s\cos\alpha \quad \text{since} \quad A = \pi s^2\sin\alpha$$

39. $\quad f(x,y) = \sqrt{x^2+y^2} \quad \text{on} \quad \Omega : 0 \le x^2+y^2 \le 1; \quad \lambda(x,y,z) = k\sqrt{x^2+y^2}$

$\quad x_M = 0, \quad y_M = 0 \qquad \text{(by symmetry)}$

$$z_M M = \iint\limits_{S} z\lambda(x,y,z)\,d\sigma = \iint\limits_{\Omega} k(x^2+y^2)\sec\left[\gamma(x,y)\right]dx\,dy$$

$$= k\sqrt{2}\iint\limits_{\Omega} (x^2+y^2)\,dx\,dy$$

$$= k\sqrt{2}\int_{0}^{2\pi}\!\!\int_{0}^{1} r^3\,dr\,d\theta = \tfrac{1}{2}\sqrt{2}\pi k$$

$\quad z_M = \tfrac{3}{4} \quad \text{since} \quad M = \tfrac{2}{3}\sqrt{2}\pi k \qquad \text{(Exercise 38)}$

41. $\quad$ no answer required

37.

$$x_M M = \iint\limits_{S} x\lambda(x,y,z)\,d\sigma = \iint\limits_{S} kx(x^2+y^2)\,d\sigma$$

$$= 2\sqrt{3}k\iint\limits_{\Omega} (u+v)\left[(u-v)^2+4u^2\right]\,du\,dv$$

$$= 2\sqrt{3}k\int_{0}^{1}\!\!\int_{0}^{1}(5u^3 - 2u^2v + uv^2 + 5u^2v - 2uv^2 + v^3)\,dv\,du$$

$$= 2\sqrt{3}\int_{0}^{1}\left(5u^3 - u^2 + \tfrac{1}{3}u + \tfrac{5}{2}u^2 - \tfrac{2}{3}u + \tfrac{1}{4}\right)\,du = \tfrac{11}{3}\sqrt{3}k$$

$\quad x_M = \tfrac{11}{9} \quad \text{since} \quad M = 3\sqrt{3}k \qquad \text{(Exercise 42)}$

45. $\quad$ Total flux out of the solid is 0. It is clear from a diagram that the outer unit normal to the cylindrical side of the solid is given by $\mathbf{n} = x\mathbf{i} + y\mathbf{j}$ in which case $\mathbf{v}\cdot\mathbf{n} = 0$. The outer unit normals to the top and bottom of the solid are $\mathbf{k}$ and $-\mathbf{k}$ respectively. So, here as well, $\mathbf{v}\cdot\mathbf{n} = 0$ and the total flux is 0.

47. $\quad$ The surface $z = \sqrt{2-(x^2+y^2)}$ is the upper half of the sphere $x^2+y^2+z^2 = 2$. The surface intersects the surface $z = x^2+y^2$ in a circle of radius 1 at height $z = 1$. Thus the upper boundary of the solid, call it S_1, is a segment of width $\sqrt{2}-1$ on a sphere of radius $\sqrt{2}$. The area of S_1 is therefore $2\pi\sqrt{2}(\sqrt{2}-1)$. (Exercise 25, Section 10.10.) The upper unit normal to S_1 is the vector

$$\mathbf{n} = \frac{1}{\sqrt{2}}(x\mathbf{i} + y\mathbf{j} + z\mathbf{k}).$$

Therefore

$$\text{flux through } S_1 = \iint_{S_1} (\mathbf{v} \cdot \mathbf{n})\, d\sigma = \frac{1}{\sqrt{2}} \iint_{S_1} \overbrace{(x^2 + y^2 + z^2)}^{2}\, d\sigma$$

$$= \sqrt{2} \iint_{S_1} d\sigma = \sqrt{2}\,(\text{area of } S_1) = 4\pi(\sqrt{2} - 1).$$

The lower boundary of the solid, call it S_2, is the graph of the function

$$f(x,y) = x^2 + y^2 \quad \text{on} \quad \Omega : 0 \le x^2 + y^2 \le 1.$$

Taking $\mathbf{n}$ as the lower unit normal, we have

$$\text{flux through } S_2 = \iint_{S_2} (\mathbf{v} \cdot \mathbf{n})\, d\sigma = \iint_{\Omega} (v_1 f_x' + v_2 f_y' - v^3)\, dx\,dy$$

$$= \iint_{\Omega} (x^2 + y^2)\, dx\,dy = \int_0^{2\pi} \int_0^1 r^3\, dr\, d\theta = \frac{1}{2}\pi.$$

The total flux out of the solid is $4\pi(\sqrt{2} - 1) + \frac{1}{2}\pi = (4\sqrt{2} - \frac{7}{2})\pi$.

SECTION 17.8

1. $\nabla \cdot \mathbf{v} = 2, \quad \nabla \times \mathbf{v} = 0$

3. $\nabla \cdot \mathbf{v} = 0, \quad \nabla \times \mathbf{v} = 0$

5. $\nabla \cdot \mathbf{v} = 6, \quad \nabla \times \mathbf{v} = 0$

7. $\nabla \cdot \mathbf{v} = yz + 1, \quad \nabla \times \mathbf{v} = -x\mathbf{i} + xy\mathbf{j} + (1 - x)z\mathbf{k}$

9. $\nabla \cdot \mathbf{v} = 1/r^2, \quad \nabla \times \mathbf{v} = 0$

11. $\nabla \cdot \mathbf{v} = 2(x + y + z)e^{r^2}, \quad \nabla \times \mathbf{v} = 2e^{r^2}\left[(y - z)\mathbf{i} - (x - z)\mathbf{j} + (x - y)\mathbf{k}\right]$

13. $\nabla \cdot \mathbf{v} = f'(x), \quad \nabla \times \mathbf{v} = 0$ 15.   use components

17. $\nabla \cdot \mathbf{F} = \dfrac{\partial P}{\partial x} + \dfrac{\partial Q}{\partial y} + \dfrac{\partial R}{\partial z} = 2 + 4 - 6 = 0$

19. $\nabla \times \mathbf{F} = \begin{vmatrix} \mathbf{i} & \mathbf{j} & \mathbf{k} \\ \dfrac{\partial}{\partial x} & \dfrac{\partial}{\partial y} & \dfrac{\partial}{\partial z} \\ x & y & -2z \end{vmatrix} = 0$

21. $\nabla^2 f = 12(x^2 + y^2 + z^2)$ 23. $\nabla^2 f = 2y^3 z^4 + 6x^2 yz^4 + 12x^2 y^3 z^2$

25. $\nabla^2 f = e^r(1 + 2r^{-1})$ 27.   (a) $2r^2$ (b) $-1/r$

29. $\nabla^2 f = \nabla^2 g(r) = \nabla \cdot (\nabla g(r)) = \nabla \cdot \left(g'(r)r^{-1}\mathbf{r}\right)$

$$= \left[(\nabla g'(r)) \cdot r^{-1}\mathbf{r}\right] + g'(r)\left(\nabla \cdot r^{-1}\mathbf{r}\right)$$

$$= \left\{\left[g''(r)r^{-1}\mathbf{r}\right] \cdot r^{-1}\mathbf{r}\right\} + g'(r)(2r^{-1})$$

$$= g''(r) + 2r^{-1}g'(r)$$

31. $\dfrac{\partial f}{\partial x} = 2x + y + 2z, \quad \dfrac{\partial^2 f}{\partial x^2} = 2; \quad \dfrac{\partial f}{\partial y} = 4y + x - 3z, \quad \dfrac{\partial^2 f}{\partial y^2} = 4;$

$\dfrac{\partial f}{\partial z} = -6z + 2x - 3y, \quad \dfrac{\partial^2 f}{\partial z^2} = -6;$

$$\dfrac{\partial^2 f}{\partial x^2} + \dfrac{\partial^2 f}{\partial y^2} + \dfrac{\partial^2 f}{\partial z^2} = 2 + 4 - 6 = 0$$

33. $n = -1$

SECTION 17.9

1. $\displaystyle\iint\limits_S (\mathbf{v} \cdot \mathbf{n})\,d\sigma = \iiint\limits_T (\nabla \cdot \mathbf{v})\,dx\,dy\,dz = \iiint\limits_T 3\,dx\,dy\,dz = 3V = 4\pi$

3. $\displaystyle\iint\limits_S (\mathbf{v} \cdot \mathbf{n})\,d\sigma = \iiint\limits_T (\nabla \cdot \mathbf{v})\,dx\,dy\,dz = \iiint\limits_T 2(x + y + z)\,dx\,dy\,dz.$

The flux is zero since the function $f(x, y, z) = 2(x + y + z)$ satisfies the relation $f(-x, -y, -z) = -f(x, y, z)$ and T is symmetric about the origin.

5.

face	$\mathbf{n}$	$\mathbf{v} \cdot \mathbf{n}$	flux	
$x = 0$	$-\mathbf{i}$	0	0	
$x = 1$	$\mathbf{i}$	1	1	
$y = 0$	$-\mathbf{j}$	0	0	total flux = 3
$y = 1$	$\mathbf{j}$	1	1	
$z = 0$	$-\mathbf{k}$	0	0	
$z = 1$	$\mathbf{k}$	1	1	

$$\iiint\limits_T (\nabla \cdot \mathbf{v})\,dx\,dy\,dz = \iiint\limits_T 3\,dx\,dy\,dz = 3V = 3$$

7.

face	n	$v \cdot n$	flux
$x = 0$	$-i$	0	0
$x = 1$	i	1	1
$y = 0$	$-j$	xz	
$y = 1$	j	$-xz$	
$z = 0$	$-k$	0	0
$z = 1$	k	1	1

fluxes add up to 0   total flux = 2

$$\iiint_T (\nabla \cdot v)\,dxdydz = \iiint_T 2\,(x+z)\,dxdydz = 2\,(\overline{x}+\overline{z})V = 2\,(\tfrac{1}{2}+\tfrac{1}{2})1 = 2$$

9. $\text{flux} = \iiint_T (1+4y+6z)\,dxdydz = (1+4\overline{y}+6\overline{z})V = (1+0+3)\,9\pi = 36\pi$

11.

$$\text{flux} = \iiint_T (2x+x-2x)\,dxdydz \quad \iiint_T x\,dxdydz$$

$$= \int_0^1 \int_0^{1-x} \int_0^{1-x-y} x\,dz\,dy\,dx$$

$$= \int_0^1 \int_0^{1-x} \left(x - x^2 - xy\right)\,dy\,dx$$

$$= \int_0^1 \left[xy - x^2 y - \frac{1}{2}xy^2\right]_0^{1-x}\,dx$$

$$= \int_0^1 \left(\frac{1}{2}x - x^2 + \frac{1}{2}x^3\right)\,dx = \frac{1}{24}$$

13.

$$\text{flux} = \iiint_T 2(x+y+z)\,dxdydz = \int_0^4 \int_0^2 \int_0^{2\pi} 2(r\cos\theta + r\sin\theta + z)r\,dr\,d\theta\,dz$$

$$= \int_0^4 \int_0^2 4\pi\,rz\,dr\,dz$$

$$= \int_0^4 8\pi\,z\,dz = 64\pi$$

15. $\text{flux} = \iiint_T (2y+2y+3y)\,dxdydz = 7\overline{y}V = 0$

17. $\text{flux} = \iiint_T (A+B+C)\,dxdydz = (A+B+C)V$

19.   Let T be the solid enclosed by S and set $\mathbf{n} = n_1\mathbf{i} + n_2\mathbf{j} + n_3\mathbf{k}$.

$$\iint\limits_{S} n_1 \, d\sigma = \iint\limits_{S} (\mathbf{i} \cdot \mathbf{n}) \, d\sigma = \iiint\limits_{T} (\nabla \cdot \mathbf{i}) \, dxdydz = \iiint\limits_{T} 0 \, dxdydz = 0.$$

Similarly

$$\iint\limits_{S} n_2 \, d\sigma = 0 \quad \text{and} \quad \iint\limits_{S} n_3 \, d\sigma = 0.$$

21.   A routine computation shows that $\nabla \cdot (\nabla f \times \nabla g) = 0$.   Therefore

$$\iint\limits_{S} [(\nabla f \times \nabla g) \cdot \mathbf{n}] \, d\sigma = \iiint\limits_{T} [\nabla \cdot (\nabla f \times \nabla g)] \, dxdydz = 0.$$

23.   Set $\mathbf{F} = F_1\mathbf{i} + F_2\mathbf{j} + F_3\mathbf{k}$.

$$F_1 = \iint\limits_{S} [\rho(z - c)\mathbf{i} \cdot \mathbf{n}] \, d\sigma = \iiint\limits_{T} [\nabla \cdot \rho(z - c)\mathbf{i}] \, dxdydz$$

$$= \iiint\limits_{T} \underbrace{\frac{\partial}{\partial x}[\rho(z - c)]}_{0} \, dxdydz = 0.$$

Similarly $F_2 = 0$.

$$F_3 = \iint\limits_{S} [\rho(z - c)\mathbf{k} \cdot \mathbf{n}] \, d\sigma = \iiint\limits_{T} [\nabla \cdot \rho(z - c)\mathbf{k}] \, dxdydz$$

$$= \iiint\limits_{T} \frac{\partial}{\partial z}[\rho(z - c)] \, dxdydz$$

$$= \iiint\limits_{T} \rho \, dxdydz = W.$$

SECTION 17.10

For Exercises 1–4: $\mathbf{n} = x\mathbf{i} + y\mathbf{j} + z\mathbf{k}$ and $C : \mathbf{r}(u) = \cos u\,\mathbf{i} + \sin u\,\mathbf{j}, \quad u \in [0, 2\pi]$.

1.   (a) $\displaystyle \iint\limits_{S} [(\nabla \times \mathbf{v}) \cdot \mathbf{n}] \, d\sigma = \iint\limits_{S} (0 \cdot \mathbf{n}) \, d\sigma = 0$

      (b) S is bounded by the unit circle $C : \mathbf{r}(u) = \cos u\,\mathbf{i} + \sin u\,\mathbf{j}, \quad u \in [0, 2\pi]$.

         $\displaystyle \oint_{C} \mathbf{v}(\mathbf{r}) \cdot d\mathbf{r} = 0$ since $\mathbf{v}$ is a gradient.

3.   (a) $\displaystyle \iint\limits_{S} [(\nabla \times \mathbf{v}) \cdot \mathbf{n}] \, d\sigma = \iint\limits_{S} [(-3y^2\mathbf{i} + 2z\mathbf{j} + 2\mathbf{k}) \cdot \mathbf{n}] \, d\sigma$

$$= \iint_S (-3xy^2 + 2yz + 2z)\, d\sigma$$

$$= \underbrace{\iint_S (-3xy^2)\, d\sigma}_{0} + \underbrace{\iint_S 2yz\, d\sigma}_{0} + 2\iint_S z\, d\sigma = 2\bar{z}V = 2(\tfrac{1}{2})2\pi = 2\pi$$

Exercise 17, Section 17.7 ⟶

(b) $\displaystyle\oint_C \mathbf{v(r)} \cdot d\mathbf{r} = \oint_C z^2\, dx + 2x\, dy = \oint_C 2x\, dy = \int_0^{2\pi} 2\cos^2 u\, du = 2\pi$

For Exercises 5–7 take $S: z = 2 - x - y$ with $0 \le x \le 2, \ 0 \le y \le 2 - x$ and C as the triangle $(2,0,0), (0,2,0), (0,0,2)$. Then $C = C_1 \cup C_2 \cup C_3$ with

$$C_1: \mathbf{r}_1(u) = 2(1-u)\mathbf{i} + 2u\mathbf{j}, \quad u \in [0,1],$$

$$C_2: \mathbf{r}_2(u) = 2(1-u)\mathbf{j} + 2u\mathbf{k}, \quad u \in [0,1],$$

$$C_3: \mathbf{r}_3(u) = 2(1-u)\mathbf{k} + 2u\mathbf{i}, \quad u \in [0,1].$$

$\mathbf{n} = \tfrac{1}{3}\sqrt{3}(\mathbf{i} + \mathbf{j} + \mathbf{k})$ area of $S: A = 2\sqrt{3}$ centroid: $\left(\tfrac{2}{3}, \tfrac{2}{3}, \tfrac{2}{3}\right)$

5. (a) $\displaystyle\iint_S [(\nabla \times \mathbf{v}) \cdot \mathbf{n}]\, d\sigma = \iint_S \tfrac{1}{3}\sqrt{3}\, d\sigma = \tfrac{1}{3}\sqrt{3}A = 2$

   (b) $\displaystyle\oint_C \mathbf{v(r)} \cdot d\mathbf{r} = \left(\int_{C_1} + \int_{C_2} + \int_{C_3}\right)\mathbf{v(r)} \cdot d\mathbf{r} = -2 + 2 + 2 = 2$

7. (a) $\displaystyle\iint_S [(\nabla \times \mathbf{v}) \cdot \mathbf{n}]\, d\sigma = \iint_S (y\mathbf{k} \cdot \mathbf{n})\, d\sigma = \tfrac{1}{3}\sqrt{3}\iint_S y\, d\sigma = \tfrac{1}{3}\sqrt{3}\bar{y}A = \tfrac{4}{3}$

   (b) $\displaystyle\oint_C \mathbf{v(r)} \cdot d\mathbf{r} = \left(\int_{C_1} + \int_{C_2} + \int_{C_3}\right)\mathbf{v(r)} \cdot d\mathbf{r} = \left(\tfrac{4}{3} - \tfrac{32}{5}\right) + \tfrac{32}{5} + 0 = \tfrac{4}{3}$

9. The bounding curve is the set of all (x, y, z) with

$$x^2 + y^2 = 4 \quad \text{and} \quad z = 4.$$

Traversed in the positive sense with respect to $\mathbf{n}$, it is the curve $-C$ where

$$C : \mathbf{r}(u) = 2\cos u\, \mathbf{i} + 2\sin u\, \mathbf{j} + 4\mathbf{k}, \qquad u \in [0, 2\pi].$$

By Stokes's theorem the flux we want is

$$-\int_C \mathbf{v(r)} \cdot \mathbf{dr} = -\int_C y\,dx + z\,dy + x^2 z^2\,dz$$

$$= -\int_0^{2\pi} \left(-4\sin^2 u + 8\cos u\right)\,du = 4\pi.$$

11. The bounding curve C for S is the bounding curve of the elliptical region $\Omega : \frac{1}{4}x^2 + \frac{1}{9}y^2 = 1$. Since

$$\nabla \times \mathbf{v} = 2x^2 yz^2 \mathbf{i} - 2xy^2 z^2 \mathbf{j}$$

is zero on the xy-plane, the flux of $\nabla \times \mathbf{v}$ through Ω is zero, the circulation of $\mathbf{v}$ about C is zero, and therefore the flux of $\nabla \times \mathbf{v}$ through S is zero.

13. C bounds the surface

$$S: z = \sqrt{1 - \tfrac{1}{2}(x^2 + y^2)}, \qquad (x, y) \in \Omega$$

with $\Omega : x^2 + (y - \frac{1}{2})^2 \le \frac{1}{4}$. Routine calculation shows that $\nabla \times \mathbf{v} = y\mathbf{k}$. The circulation of $\mathbf{v}$ with respect to the upper unit normal $\mathbf{n}$ is given by

$$\iint_S (y\mathbf{k} \cdot \mathbf{n})\,d\sigma = \iint_\Omega y\,dxdy = \overline{y}A = \frac{1}{2}\left(\frac{\pi}{4}\right) = \frac{1}{8}\pi.$$
$$(17.7.9)$$

If $-\mathbf{n}$ is used, the circulation is $-\frac{1}{8}\pi$.   Answer: $\pm\frac{1}{8}\pi$.

15. $\nabla \times \mathbf{v} = \mathbf{i} + 2\mathbf{j} + \mathbf{k}$. The paraboloid intersects the plane in a curve C that bounds a flat surface S that projects onto the disc $x^2 + (y - \frac{1}{2})^2 = \frac{1}{4}$ in the xy-plane. The upper unit normal to S is the vector $\mathbf{n} = \frac{1}{2}\sqrt{2}\,(-\mathbf{j} + \mathbf{k})$. The area of the base disc is $\frac{1}{4}\pi$. Letting γ be the angle between $\mathbf{n}$ and $\mathbf{k}$, we have $\cos\gamma = \mathbf{n} \cdot \mathbf{k} = \frac{1}{2}\sqrt{2}$ and $\sec\gamma = \sqrt{2}$. Therefore the area of S is $\frac{1}{4}\sqrt{2}\pi$. The circulation of $\mathbf{v}$ with respect to $\mathbf{n}$ is given by

$$\iint_S [(\nabla \times \mathbf{v}) \cdot \mathbf{n}]\,d\sigma = \iint_S -\frac{1}{2}\sqrt{2}\,d\sigma = \left(-\frac{1}{2}\sqrt{2}\right)(\text{area of } S) = -\frac{1}{4}\pi.$$

If $-\mathbf{n}$ is used, the circulation is $\frac{1}{4}\pi$.   Answer: $\pm\frac{1}{4}\pi$.

17. Straightforward calculation shows that

$$\nabla \times (\mathbf{a} \times \mathbf{r}) = \nabla \times [(a_2 z - a_3 y)\mathbf{i} + (a_3 x - a_1 z)\mathbf{j} + (a_1 y - a_2 x)\mathbf{k}] = 2\mathbf{a}.$$

19. In the plane of C, the curve C bounds some Jordan region that we call Ω. The surface $S \cup \Omega$ is a piecewise–smooth surface that bounds a solid T. Note that $\nabla \times \mathbf{v}$ is continuously differentiable on T. Thus, by the divergence theorem,

$$\iiint_T [\nabla \cdot (\nabla \times \mathbf{v})]\, dx\,dy\,dz = \iint_{S \cup \Omega} [(\nabla \times \mathbf{v}) \cdot \mathbf{n}]\, d\sigma$$

where $\mathbf{n}$ is the outer unit normal. Since the divergence of a curl is identically zero, we have

$$\iint_{S \cup \Omega} [(\nabla \times \mathbf{v}) \cdot \mathbf{n}]\, d\sigma = 0.$$

Now $\mathbf{n}$ is $\mathbf{n}_1$ on S and $\mathbf{n}_2$ on Ω. Thus

$$\iint_S [(\nabla \times \mathbf{v}) \cdot \mathbf{n}_1]\, d\sigma + \iint_\Omega [(\nabla \times \mathbf{v}) \cdot \mathbf{n}_2]\, d\sigma = 0.$$

This gives

$$\iint_S [(\nabla \times \mathbf{v}) \cdot \mathbf{n}_1]\, d\sigma = \iint_\Omega [(\nabla \times \mathbf{v}) \cdot (-\mathbf{n}_2)]\, d\sigma = \oint_C \mathbf{v}(\mathbf{r}) \cdot d\mathbf{r}$$

where C is traversed in a positive sense with respect to $-\mathbf{n}_2$ and therefore in a positive sense with respect to $\mathbf{n}_1$. ($-\mathbf{n}_2$ points toward S.)

PROJECTS AND EXPLORATIONS

17.1.   (a) For $\mathbf{f}(x,y) = y\,\mathbf{i} + x\,\mathbf{j}$ and $\mathbf{r}(u) = u\,\mathbf{i} + u^2\,\mathbf{j}$, $u \in [0,1]$,

$$\int_C \mathbf{f}(\mathbf{r}) \cdot d\mathbf{r} = \int_0^1 3u^2\, du = [u^3]_0^1 = 1.$$

For $\mathbf{f}(x,y,z) = y^2 z^3\,\mathbf{i} + 2xyz^3\,\mathbf{j} + 3xy^2 z^2\,\mathbf{k}$ and $\mathbf{r}(u) = u^2\,\mathbf{i} + u^4\,\mathbf{j}, + u^6\,\mathbf{k}$ $u \in [0,1]$,

$$\int_C \mathbf{f}(\mathbf{r}) \cdot d\mathbf{r} = \int_0^1 283u^{27}\, du = [u^{28}]_0^1 = 1.$$

(d) The arc length of this curve is approximately 7.223048. Most of the exact or approximate methods for calculating the arc length as a line integral have problems since the derivative of the $\mathbf{k}$ -term is not defined at $u = \pi$. One alternative for (i) is:

$\mathbf{r}(u) = u \sin u\,\mathbf{i} + u \cos u\,\mathbf{j} + \sqrt{\pi^2 - u^2}\,\mathbf{k}$, $u \in [0, \pi/2]$;

$\mathbf{r}(v) = \sqrt{\pi^2 - v^2} \sin \sqrt{\pi^2 - v^2}\,\mathbf{i} + \sqrt{\pi^2 - v^2} \cos \sqrt{\pi^2 - v^2}\,\mathbf{j} + v\,\mathbf{k}$ $u \in [0, \sqrt{3}\pi/2]$.

A similar problem occurs with $\mathbf{R}(u)$.

(e) The value of the line integrals is approximately 17.874418.

(f) We want the points randomly distributed along the curve which is not the same as being randomly distributed over the parameter interval.

17.3.   (a) For $\mathbf{v}(x, y, z) = xyz\,\mathbf{i} + \ln(xyz)\,\mathbf{j} - \sin(2x + 3y - 4z)\,\mathbf{k}$:

$$\nabla \cdot \mathbf{v} = yz + \frac{1}{y} + 4\,\cos(2x + 3y - 4z);$$

$$\nabla \times \mathbf{v} = \left[-3\,\cos(2x + 3y - 4z) - \frac{1}{z}\right]\mathbf{i} + \left[xy + 2\,\cos(2x + 3y - 4z)\right]\mathbf{j} + \left[\frac{1}{x} - xz\right]\mathbf{k}.$$

For $\mathbf{v}(x, y, z) = \left(x^2 + y^3 - 3z^7\right)\mathbf{i} - \tan^{-1}(x/y)\,\mathbf{j} + e^{x-z}\,\mathbf{k}$:

$$\nabla \cdot \mathbf{v} = 2x + \frac{x}{x^2 + y^2} - e^{x-z};$$

$$\nabla \times \mathbf{v} = \left[e^{x-z} - 21z^6\right]\mathbf{j} - \left[\frac{y}{x^2 + y^2} + 3y^2\right]\mathbf{k}.$$

For $f(x, y, z) = \cos\sqrt{x^2 + 3\,\sin y + 5z^4}$, let $G = \sqrt{x^2 + 3\,\sin y + 5z^4}$. Then

$$\nabla f = \frac{-x}{G}\,\sin G\,\mathbf{i} - \frac{3\,\cos y}{2G}\,\sin G\,\mathbf{j} - \frac{10x^3}{G}\,\sin G\,\mathbf{k}.$$

(c) See Exercise 13, Section 17.10.

FOR SOLUTIONS TO CHAPTER 18:
ELEMENTARY DIFFERENTIAL EQUATIONS
PLEASE REFER TO CHAPTER 12 IN THE
STUDENT SOLUTIONS MANUAL
TO ACCOMPANY
SALAS AND HILLE'S
CALCULUS, ONE VARIABLE
EARLY TRANSCENDENTALS, 7E